Frank Pörner

Regularization Methods for Ill-Posed Optimal Control Problems

Frank Pörner

Regularization Methods for Ill-Posed Optimal Control Problems

Würzburg
University Press

Dissertation, Julius-Maximilians-Universität Würzburg
Fakultät für Mathematik und Informatik, 2018
Gutachter: Prof. Dr. Daniel Wachsmuth, Prof. Dr. Christian Clason

Impressum

Julius-Maximilians-Universität Würzburg
Würzburg University Press
Universitätsbibliothek Würzburg
Am Hubland
D-97074 Würzburg
www.wup.uni-wuerzburg.de

© 2018 Würzburg University Press
Print on Demand

Coverdesign: Julia Bauer

ISBN 978-3-95826-086-3 (print)
ISBN 978-3-95826-087-0 (online)
URN urn:nbn:de:bvb:20-opus-163153

Preface

Abstract

This thesis deals with the construction and analysis of solution methods for a class of ill-posed optimal control problems involving elliptic partial differential equations as well as inequality constraints for the control and state variables. The objective functional is of tracking type, without any additional L^2-regularization terms. This makes the problem ill-posed and numerically challenging.

We split this thesis in two parts. The first part deals with linear elliptic partial differential equations. In this case, the resulting solution operator of the partial differential equation is linear, making the objective functional linear-quadratic. To cope with additional control constraints we introduce and analyse an iterative regularization method based on Bregman distances. This method reduces to the proximal point method for a specific choice of the regularization functional. It turns out that this is an efficient method for the solution of ill-posed optimal control problems. We derive regularization error estimates under a regularity assumption which is a combination of a source condition and a structural assumption on the active sets. If additional state constraints are present we combine an augmented Lagrange approach with a Tikhonov regularization scheme to solve this problem.

The second part deals with non-linear elliptic partial differential equations. This significantly increases the complexity of the optimal control as the associated solution operator of the partial differential equation is now non-linear. In order to regularize and solve this problem we apply a Tikhonov regularization method and analyse this problem with the help of a suitable second order condition. Regularization error estimates are again derived under a regularity assumption. These results are then extended to a sparsity promoting objective functional.

Zusammenfassung

Diese Arbeit beschäftigt sich mit der Konstruktion und Analyse von Lösungsverfahren für schlecht gestellte Steuerungsprobleme. Die Nebenbedingungen sind in der Form von elliptischen partiellen Differentialgleichungen, sowie Ungleichungsrestriktionen für die Steuerung und den zugehörigen Zustand gegeben. Das Zielfunktional besteht aus einem Tracking-Type-Term ohne zusätzliche L^2-Regularisierungsterme. Dies führt dazu, dass das Optimalsteuerungsproblem schlecht gestellt ist, was die numerische Berechnung einer Lösung erschwert.

Diese Arbeit ist in zwei Teile aufgeteilt. Der erste Teil beschäftigt sich mit linearen elliptischen partiellen Differentialgleichungen. In diesem Fall ist der zugehörige Lösungsoperator der partiellen Differentialgleichung linear und das Zielfunktional linear-quadratisch. Um die zusätzlichen Steuerungsrestriktionen zu behandeln, betrachten wir ein iteratives Verfahren welches auf einer Regularisierung mit Bregman-Abständen basiert. Für eine spezielle Wahl des Regularisierungsfunktionals vereinfacht sich dieses Verfahren zu dem Proximal-Point-Verfahren. Die Analyse des Verfahrens zeigt, dass es ein effizientes und gut geeignetes Verfahren ist, um schlecht gestellte Optimalsteuerungsprobleme zu lösen. Mithilfe einer Regularitätsannahme werden Konvergenzraten für den Regularisierungsfehler hergeleitet. Diese Regularitätsannahme ist eine Kombination einer Source-Condition sowie einer struktuellen Annahme an die aktiven Mengen. Wenn zusätzlich Zustandsrestriktionen vorhanden sind, wird zur Lösung eine Kombination aus dem Augmented Lagrange Ansatz sowie einer Tikhonov-Regularisierung angewendet.

Der zweite Teil dieser Arbeit betrachtet nicht-lineare partielle Differentialgleichungen. Dies erhöht die Komplexität des Optimalsteuerungsproblem signifikant, da der Lösungsoperator der partiellen Differentialgleichung nun nicht-linear ist. Zur Lösung wird eine Tikhonov-Regularisierung betrachtet. Mithilfe einer geeigneten Bedingung zweiter Ordnung wird dieses Verfahren analysiert. Auch hier werden Konvergenzraten mithilfe einer Regularitätsannahme bestimmt. Anschließend werden diese Methoden auf ein Funktional mit einem zusätzlichen L^1-Term angewendet.

Acknowledgments

I would first of all like to thank my supervisor Prof. Dr. Daniel Wachsmuth for offering me a doctoral position, supervising my thesis and supporting me throughout the last years. He introduced my to the topic of optimal control of partial differential equations and related topics. From the very beginning of my work he was always willing to discuss any problems and question and gave many valuable ideas.

At the same time I want to express my appreciation to Prof. Dr. Christian Clason who agreed to co-referee this thesis.

Moreover I would like to thank all members of the institute of mathematics of the Julius-Maximilians-Universität Würzburg, in particular the chair of optimization and numerical mathematics lead by Prof. Dr. Christian Kanzow. I enjoyed the scientific atmosphere which led to this thesis and the submission of five scientific papers.

Special thanks go to my colleagues Veronika Karl and Daniel Steck for several fruitful discussions - not only about mathematics. Furthermore I really enjoyed several games of chess with Daniel Steck - unfortunately I was not able to win a single game.

Furthermore, I am also grateful for the support of my family and friends who supported and encouraged me during my studies.

Last, but not least I want to acknowledge the financial support by the German Research Foundation DFG under the grant Wa 3626/1-1.

Frank Pörner,
Würzburg, August 2018

List of Symbols and Abbreviations

General Notation

$\exists$	There exists
$\forall$	For all
$\mathbb{N}$	The set of all natural numbers $\{1, 2, 3, ...\}$
$\mathbb{N}_0$	The set of all natural numbers with 0
$\mathbb{R}$	The set of all real numbers
$\mathbb{R}^n$	The n-dimensional vector space of real numbers
$\overline{\mathbb{R}}$	The extended real numbers $\mathbb{R} \cup \{-\infty, +\infty\}$
$\mathrm{meas}(M)$	The Lebesgue measure of a measurable set M

Partial Differential Equations

Ω	An open, bounded domain of $\mathbb{R}^n$
$\partial\Omega$	The boundary of Ω
D^α	The differential operator with respect to the multiindex $\alpha \in \mathbb{N}_0^n$
∇	The differential operator $(\partial_{x_i})_{i=1,...,n}$ as column vector
Δv	The Laplacian $\nabla \cdot \nabla v$

Functional Analysis

$C(\Omega)$	The vector space of continuous functions with image in $\mathbb{R}$
$C(\overline{\Omega})$	Functions in $C(\Omega)$ which continuously extends to $\partial\Omega$
$L^p(\Omega)$	The Banach space of p-times Lebesgue-integrable functions
$L^\infty(\Omega)$	The Banach space of essentially bounded functions
$W^{k,p}(\Omega)$	Sobolev space of order k
$H^k(\Omega)$	The space $W^{k,2}(\Omega)$
$H_0^1(\Omega)$	The set of all functions $v \in H^1(\Omega)$ with $v = 0$ on $\partial\Omega$
$\|\cdot\|_V$	The norm in the vector space V
$\mathcal{L}(V, W)$	The set of linear and bounded operators from V to W
V^*	The dual space $\mathcal{L}(V, \mathbb{R})$ of V
$(\cdot, \cdot)_H$	The inner product in a Hilbert space H
$\langle \cdot, \cdot \rangle_{V^*, V}$	The duality pairing of elements from V^*, V
$\rightarrow$	Strong convergence
$\rightharpoonup$	Weak convergence
$\overset{*}{\rightharpoonup}$	Weak-star convergence

Contents

Preface   v

List of Symbols and Abbreviations   ix

1 Introduction   1

2 Preliminaries   5
   2.1 Tools from Functional Analysis 5
      2.1.1 Functional Analysis 5
      2.1.2 Function Spaces 8
   2.2 Ill-Posedness and Regularization Techniques 10
      2.2.1 Ill-Posedness of Optimal Control Problems 10
      2.2.2 Poisson's Equation 11
      2.2.3 The Tikhonov Regularization 12
      2.2.4 The Proximal Point Method 14
      2.2.5 The Iterative Bregman Method 14
      2.2.6 The Augmented Lagrange Method 15

I Linear State Equation   17

3 Iterative Bregman Method for Control Constraints   19
   3.1 The Linear-Quadratic Problem 20
   3.2 Optimality Conditions 20
   3.3 Regularity Assumption 21
      3.3.1 Source Condition 21
      3.3.2 Active Set Condition 22
   3.4 The Bregman Iteration 26
      3.4.1 Bregman Distance 27
      3.4.2 The Iterative Bregman Method 28
      3.4.3 A-Priori Error Estimates 31
      3.4.4 Auxiliary Estimates 32
      3.4.5 Convergence Results 35
      3.4.6 Noise Estimates 44
      3.4.7 A-Priori Stopping Rule 48

4   The Inexact Iterative Bregman Method   **53**
  4.1   The Discretized Problem   53
      4.1.1   The Operator S_h   53
      4.1.2   A-Posteriori Error Estimate for the Discretized Subproblem   55
  4.2   Inexact Bregman Iteration   56
      4.2.1   Notation and Auxiliary Results   57
      4.2.2   Convergence under Source Condition   58
      4.2.3   Convergence under Active Set Condition   63

5   Implementation of the Iterative Bregman Method   **69**
  5.1   Semi-Smooth Newton Method   69
      5.1.1   Algorithmic Aspects and Implementation   71
      5.1.2   Using the Newton Solver in the Bregman Iteration   73
  5.2   Numerical Results   74
      5.2.1   One-Dimensional Examples   74
      5.2.2   Two-Dimensional Examples   79

6   ALM for State Constraints and Sparsity   **85**
  6.1   Model Problem with State Constraints   86
  6.2   Preliminary Results   88
      6.2.1   Problem Setting   88
      6.2.2   Subdifferential of the L^1-Norm   89
      6.2.3   Optimality Conditions   90
  6.3   Convergence Analysis of the Regularized Problem   92
  6.4   The Augmented Lagrange Method   97
      6.4.1   The Augmented Lagrange Optimal Control Problem   97
      6.4.2   The Prototypical Augmented Lagrange Algorithm   99
      6.4.3   The Multiplier Update Rule   100
      6.4.4   The Augmented Lagrange Algorithm in Detail   101
      6.4.5   Infinitely Many Successful Steps   102
  6.5   Convergence Results   103

7   Numerical Implementation of the ALM   **109**
  7.1   Active-Set Method   110
  7.2   Numerical Results   113
      7.2.1   Bang-Bang-Off Solution in One Space Dimension   114
      7.2.2   Bang-Bang-Off Solution in Two Space Dimensions   116
      7.2.3   Bang-Bang-Off Solution on Unit Square   118

II   Nonlinear State Equation   **121**

8   Tikhonov regularization   **123**
  8.1   Problem Setting   124
  8.2   Assumptions and Preliminary Results   125
      8.2.1   Existence of Solutions   126

8.2.2 Second-Order Optimality Conditions 128
8.2.3 Regularity Conditions 131
8.3 Convergence Results 132
8.3.1 Analysis of the Tikhonov Regularization 132
8.3.2 Convergence Rates 134
8.4 Necessity of the Regularity Assumption 139
8.5 Numerical Results 141

9 Extension to Sparse Control Problems   **145**
9.1 Model Problem with Sparsity 145
9.2 The Tikhonov Regularization 146
9.3 Sufficient Second Order Conditions 147
9.4 Regularity Assumption 147
9.5 Convergence Rates 150

10 Conclusion and Outlook   **155**
10.1 Linear State Equation 155
10.1.1 Control Constraints 155
10.1.2 Control and State Constraints 156
10.2 Nonlinear State Equation 157

Bibliography   **159**

CHAPTER 1

Introduction

Mathematics is a powerful tool to model phenomena and problems in a very wide range of all scientific and industrial areas. Especially in physics the concept of partial differential equations (PDE) is an excellent tool to describe many physical problems. For instance, these equations can be used to model many problems arising in fluid dynamics, let us here mention the Euler and magnetohydrodynamics equations as very popular examples. Further examples which can be modelled by partial differential equations are heat conduction, electric potentials and many biological processes [69].

In many physical models and systems, the role of the PDE is that it maps a certain initial state to the resulting output. We will refer to this problem as the forward equation. Most classes of partial differential equations are deterministic, i.e. for a given initial state it will always generate the same output. However, in many physical situations it is not possible to observe or measure the initial state. It is only possible to measure the output. To reconstruct the associated initial state one has to reverse the forward equation in a certain sense. This can be formulated as an PDE-constrained optimal control problem.

In the context of such optimal control problem the initial state is called control u and the output generated by the PDE is referred as state y. The PDE is described by the operator S, hence we have $Su = y$. The measured output will be called desired state y_d. We consider a so called tracking type functional, hence we want to steer the control u in such a way that the associated state y is as close as possible to y_d. This can be formulated as the constrained minimization problem

$$\min_{u,y} \ \frac{1}{2}\|y - y_d\|^2,$$
$$\text{such that} \quad Su = y.$$

From another point of view, the desired state y_d can be interpreted as a given value and we want to control the system such that the output is again as close as possible to the desired state. An example is the optimal cooling or heating of a manufactured component.

In many physical models there are additional constraints for the control and state. Consider again the optimal heating problem. Due to temperature restrictions for the heat source we have to bound the control. This can be formulated as the constrained minimization problem

$$\min_{u,y} \quad \frac{1}{2}\|y - y_d\|^2,$$
$$\text{such that} \quad Su = y, \tag{1.1}$$
$$u_a \le u \le u_b,$$

with $u_a \le u_b$. In many situations the operator S is linear and compact which makes (1.1) ill-posed and numerically difficult to solve as small errors in the desired state may lead to a big error in the optimal control. A remedy is to apply different regularization techniques. We want to mention the Tikhonov regularization [33, 93, 96, 97, 99], the iterated Tikhonov or proximal point method [33, 42, 45, 56, 67, 79, 80, 82, 83, 88] and the iterative Bregman method [9–11, 21–23, 30–32, 38, 72] and the references therein.

We use the iterative Bregman method to solve the constrained minimzation problem (1.1). We analyse this method with respect to convergence and numerical stability under a regularity assumption which includes possible non-attainability and bang-bang solutions. This regularity condition is a combination of a classical source condition and a structural assumption on the active sets [28, 74–78, 97, 99].

We establish regularization error estimates for control, state and adjoint state. Furthermore we establish a stopping rule for noisy data and analyse an inexact version of the iterative Bregman method.

However, in many situations additional constraints for the state have to be imposed. In the case of the optimal heating a limitation of the state is needed to avoid mechanical damage through overheating. This can be modelled by the additional state constraints $y \le \psi$. Hence we are interested in the solution of the problem

$$\min_{u,y} \quad \frac{1}{2}\|y - y_d\|^2 + \beta\|u\|_{L^1(\Omega)},$$
$$\text{such that} \quad Su = y, \tag{1.2}$$
$$y \le \psi,$$
$$u_a \le u \le u_b.$$

The additional L^1-term guarantees that the resulting optimal control is sparse, i.e. it is zero on large parts of Ω. Starting with the pioneering work [89] such sparsity promoting functionals have been studied in [97–99] but without additional state constraints. These additional state constraint significantly increases the complexity of the problem. A possible approach to handle the constraint is to use the augmented Lagrange method [2, 3, 50, 55, 59]. We couple the Tikhonov regularization scheme with the augmented Lagrange approach and couple the regularization parameter with the penalty parameter to obtain a stable scheme. We introduce an update rule for the multiplier which allows us to prove convergence of the method.

Up to now the operator S associated with the forward equation was assumed to be linear. Although many physical problems can be modelled with a linear PDE in some cases a more sophisticated PDE is needed. We therefore also consider problems of the form

$$\min_{u,y} \quad \frac{1}{2}\|y - y_d\|^2 + \beta\|u\|_{L^1(\Omega)},$$
$$\text{such that} \quad Su = y, \tag{1.3}$$
$$u_a \le u \le u_b,$$

which resembles (1.1) but now S is a non-linear operator. To be precise we now assume that the underlying PDE is semi-linear. Such sparsity promoting functionals for a non-linear PDE have been studied in [13, 16]. Optimal control of semi-linear partial differential equations has been intensively studied in the literature, see [7, 13, 15, 16, 19, 20, 33, 63, 71, 93] and the references therein.

We extend the regularity condition used for the Bregman iteration to the non-linear case. As the problem (1.3) is non-convex we make heavy use of a second order sufficient condition presented by Casas [13] and our regularity assumption to prove convergence rates of a Tikhonov regularization of (1.3). To the best of our knowledge this is the first convergence rate result subject to non-linear partial differential equations. Furthermore we show that our regularity assumption is not only sufficient but also necessary for high convergence rates.

This thesis is organized as follows. In Chapter 2 we introduce the mathematical tools and concepts needed in this thesis. In particular we start with the functional analytic preliminaries and introduce the concepts of the regularization methods mentioned in the introduction.

Chapter 3 is devoted to the iterative Bregman method to solve (1.1). We introduce our regularity condition and the Bregman method. We provide regularization error estimates and an a-priori stopping rule for noisy data. In Chapter 4 we introduce an inexact version of the iterative Bregman method. We describe the discretization and establish convergence and stability results. Chapter 5 deals with the numerical and practical implementation of the iterative Bregman method. We establish a semi-smooth Newton method and present several test examples.

In Chapter 6 we introduce the augmented Lagrange method for (1.2). We show how we couple the Tikhonov regularization and the augmented Lagrange approach and present convergence results. Chapter 7 shows how to implement the augmented Lagrange method presented in Chapter 6 using an active-set method.

In Chapter 8 we consider (1.3) with a non-linear partial differential equation and $\beta = 0$. We transfer the regularity condition to the non-linear case and introduce second order conditions. By combining both assumptions we establish convergence rates and show that the regularity assumption is not only sufficient but also necessary for high convergence rates. In Chapter 9 we introduce second order conditions and transfer our regularity condition to (1.3) with $\beta > 0$. Again convergence results are presented.

Finally in Chapter 10 we summarize our results and list some possible extensions and modifications for future research.

CHAPTER 2

Preliminaries

2.1 Tools from Functional Analysis

The aim of the this section is to collect and provide the necessary tools from functional analysis and optimization which we need during this thesis. We do not provide any proofs for the results presented in the following. Instead we want to refer to the books of Werner [100], Dobrowolski [29], Tröltzsch [93], Bonnans and Shapiro [8] and Rudin [84].

2.1.1 Functional Analysis

Let V, W be two real Banach spaces with norms $\|\cdot\|_V$ and $\|\cdot\|_W$, respectively. An operator $A : V \to W$ is called bounded if there exists a constant $c \geq 0$ such that $\|Av\|_W \leq c\|v\|_V$ for all $v \in V$. We denote $\mathcal{L}(V, W)$ the set of all linear and bounded operators between V and W. This set endowed with the norm

$$\|A\|_{\mathcal{L}(V,W)} := \inf\{c \geq 0 \mid \forall x \in V : \|Av\|_W \leq c\|x\|_V\}$$

is a Banach space itself. For the case $V \neq \{0\}$ this is equivalent to

$$\|A\|_{\mathcal{L}(V,W)} = \sup_{\|v\|_V = 1} \|Av\|_W.$$

For linear operators the concepts of boundedness and continuity coincide.

For every Banach space V the dual space V^* is defined by $V^* := \mathcal{L}(V, \mathbb{R})$. We say that the space V is reflexive if the canonical embedding

$$i : V \to (V^*)^*, \quad v \mapsto [v' \in V^* \mapsto v'(v)]$$

is bijective. In this case we identify $(V^*)^* = V$. We define the duality pairing

$$\langle v', v \rangle_{V^*,V} := v'(v), \quad v' \in V^*, \ v \in V.$$

We drop the subscript V^*, V if the spaces V and V^* are clear from the context. We say a sequence $(v_k)_k \subset V$ converges weakly to $v \in V$ if and only if

$$\lim_{k \to \infty} \langle v', v_k \rangle_{V^*, V} = \langle v', v \rangle_{V^*, V} \quad \forall \tilde{v}' \in V^*.$$

A common abbreviation in the literature is to write $v_k \rightharpoonup v$ if $(v_k)_k$ converges weakly to v. Furthermore we say that $(v_k')_k \subset V^*$ converges weakly* to $v' \in V^*$ if and only if

$$\lim_{k \to \infty} v_k'(v) = v'(v) \quad \forall v \in V.$$

In this case we write $v_k' \rightharpoonup^* v'$ in short. Note that if V is reflexive, weak and weak* convergence coincide in V^*.

If the norm $\| \cdot \|_H$ of a Banach space H is induced by an inner product, we speak of a Hilbert space with inner product $(\cdot, \cdot)_H$. For a Hilbert space every linear and bounded operator can be characterized. This is part of the next result, which is known as the Riesz representation theorem.

Theorem 2.1.1 (Riesz). *Let H be a real Hilbert space. For all $l' \in H^*$, there exists an element $l \in H$ such that $\langle l', v \rangle_{H^*, H} = (l, v)_H$ for all $v \in H$, and $\|l'\|_{H^*} = \|l\|_H$.*

Another well-known theorem is the Lax-Milgram theorem, which is a powerful tool to analyse partial differential equations.

Theorem 2.1.2 (Lax-Milgram). *Let H be a real Hilbert space, $f \in H^*$ and $a : H \times H \to \mathbb{R}$ a bilinear mapping which is coercive, i.e. there exists a constant $c_1 > 0$ such that*

$$a(y, y) \geq c_1 \|y\|_H^2 \quad \forall y \in H,$$

and bounded, i.e. there exists a constant $c_2 > 0$ such that

$$|a(y, v)| \leq c_2 \|y\|_H \|v\|_H \quad \forall y, v \in H.$$

Then, there exists a uniquely determined $y \in H$ such that

$$a(y, v) = f(v) \quad \forall v \in H$$

holds. Furthermore, the a-priori bound $\|y\|_H \leq c_1^{-1} \|f\|_{H^*}$ holds.

Let us now introduce the adjoint operator. For $w' \in W^*$ and a linear operator $A : V \to W$, the adjoint operator $A^* : W^* \to V^*$ is defined by

$$(A^* w') v := w'(Av).$$

Using the duality pairing introduced above we can now write

$$\langle A^* w', v \rangle_{V^*, V} = \langle w', Av \rangle_{W^*, W} \quad \forall w' \in W^*, v \in V.$$

The mapping of an operator to its adjoint is linear and isometric. We are interested in the special case $V = W$ being a Hilbert space. In this case we call A self-adjoint if and only if $(A^*)^* = A$.

A operator $A : X \to Y$ between two Banach spaces is called compact if it is continuous and maps every bounded subset of X into a relatively compact set of Y, i.e. its closure is compact. As a immediate result A maps weakly converging sequences to strongly converging sequences.

Let $A : X \to Y$ be a linear compact operator and X, Y be infinite dimensional Banach spaces, then A is not bijective and therefore A^{-1} cannot exists. Hence, an injective compact operator is not surjective. If $A \in \mathcal{L}(X, Y)$ is compact its adjoint operator A^* is also compact. Denote $\mathcal{R}(A)$ the range of A. Then $\mathcal{R}(A)$ is closed if and only if $\dim(\mathcal{R}(A)) < \infty$.

In the following let V be a real Banach space and denote $\bar{\mathbb{R}} := \mathbb{R} \cup \{-\infty, \infty\}$ the extended real numbers. Let $f : V \to \bar{\mathbb{R}}$ be a function. The function f is called subdifferentiable at $x \in V$, if $f(x)$ is finite and there exists $g \in V^*$ such that

$$f(y) - f(x) \geq \langle g, y - x \rangle_{V^*, V} \quad \forall y \in V.$$

The element g is called subgradient of f at x. The set of all subgradients is called subdifferential

$$\partial f(x) := \{g \in V^* : f(y) - f(x) \geq \langle g, y - x \rangle_{V^*, V} \quad \forall y \in V\}.$$

Note that the subdifferential may be empty. However, for convex functions we have the following result.

Theorem 2.1.3. *Let $f : V \to \bar{\mathbb{R}}$ be a convex function. Let $x \in V$ such that f is finite and continuous at x. Then $\partial f(x) \neq \emptyset$.*

For the subdifferential a sum rule can be established, see [86, Proposition 4.5.1].

Lemma 2.1.4. *Let $f_1, f_2 : V \to \bar{\mathbb{R}}$ proper and convex. Assume there exists an element $\bar{x} \in (\text{dom } f_1) \cap (\text{int dom } f_2)$ such that f_2 is continuous at $\bar{x}$. Then for each $x \in (\text{dom } f_1) \cap (\text{dom } f_2)$ it holds*

$$\partial(f_1 + f_2)(x) = \partial f_1(x) + \partial f_2(x).$$

Let us present two important examples which will be used during this thesis. For the first example let $C \subset V$ be a given convex set and I_C its associated indicator function. Then it holds for $x \in C$

$$\partial I_C(x) = \{x^* \in V^* : \langle x^*, y - x \rangle_{V^*, V} \leq 0, \ \forall y \in C\} = N_C(x),$$

which is the normal cone of C at x. For $x \notin C$ the subdifferential is empty, since for $y \in C$ there exists no $g \in V^*$ such that

$$0 = I_C(y) \geq \overbrace{I_C(x)}^{=+\infty} + \langle g, y - x \rangle_{V^*, V}.$$

Now let H be a real Hilbert space and define $f(x) := \frac{1}{2} \|x\|_H^2$. Define the functional $g_x : H \to \mathbb{R}, \ g_x(v) := (x, v)_H$. Then it holds for all $x \in H$

$$\partial f(x) = \{g_x\},$$

hence the subdifferential is a singleton. A direct consequence of the definition of the subdifferential is the next result.

Theorem 2.1.5. *Let $f : V \to \mathbb{R}$ be a function. Then $x \in V$ is a minimizer of f if and only if $0 \in \partial f(x)$.*

Furthermore the subdifferential can be used to derive necessary and sufficient first order conditions. The next theorem is taken from [86, Proposition 5.2.5] and is a direct consequence of Lemma 2.1.4 and Theorem 2.1.5.

Theorem 2.1.6. *Let $f : V \to \mathbb{R}$ be a continuous and convex function and $A \subset V$ nonempty and convex. Let $\bar{x} \in A$. Then the following statements are equivalent:*

1. *The element $\bar{x}$ is a solution of $\min\limits_{x \in A} f(x)$.*

2. *There exists $x^* \in \partial f(\bar{x})$ such that $\langle x^*, x - \bar{x} \rangle_{V^*, V} \geq 0 \quad \forall x \in A$.*

3. *The directional derivative satisfies $f'(\bar{x}; x - \bar{x}) \geq 0 \quad \forall x \in A$.*

We will also make use of the following theorem.

Theorem 2.1.7. *The subdifferential is monotone, i.e. for $x_1, x_2 \in V$ and $x_i^* \in \partial f(x_i)$ for $i = 1, 2$ we get*

$$\langle x_1^* - x_2^*, x_1 - x_2 \rangle_{V^*, V} \geq 0.$$

2.1.2 Function Spaces

In general, a domain Ω is an open and connected subset of $\mathbb{R}^n$ with $n \in \mathbb{N}$. In this work we will only consider bounded domains. Now let $1 \leq p < \infty$. By $L^p(\Omega)$ we denote the space of functions whose p-th power is integrable. The case $p = \infty$ is treated separately: essentially bounded functions are collected in the set $L^\infty(\Omega)$. We always restrict ourselves to the Lebesgue-measure, which will be denoted with meas.

The set $L^p(\Omega)$ consists of equivalence classes of functions. For $u, v \in L^p(\Omega)$ we say $u = v$ if $u(x) = v(x)$ holds almost everywhere (a.e.) in Ω, i.e. $u(x) \neq v(x)$ only on a set of measure zero.

Endowed with the norms

$$\|v\|_{L^p(\Omega)} := \left(\int_\Omega |v(x)|^p \right)^{1/p} \quad \text{for } 1 \leq p < \infty,$$

$$\|v\|_{L^p(\Omega)} := \operatorname*{ess\,sup}_{x \in \Omega} |v(x)| \quad \text{for } p = \infty,$$

the set $L^p(\Omega)$ becomes a Banach space. For the special case $p = 2$ we define the inner product

$$(v, w)_{L^2(\Omega)} := \int_\Omega v(x) w(x) \, \mathrm{d}x,$$

to make $L^2(\Omega)$ a Hilbert space. In this case we have $\|v\|_{L^2(\Omega)} = \sqrt{(v, v)_{L^2(\Omega)}}$. In the following we suppress the spatial argument x, unless it is explicitly needed. Let us now collect some important inequalities for L^p-functions.

For $p, q \in (1, \infty)$ with $p^{-1} + q^{-1} = 1$ and $v \in L^p(\Omega)$ and $w \in L^q(\Omega)$ it holds $vw \in L^1(\Omega)$ and

$$\int_\Omega vw \; dx \leq \left(\int_\Omega |v|^p \; dx \right)^{1/p} \left(\int_\Omega |w|^q \; dx \right)^{1/q} = \|v\|_{L^p(\Omega)} \|w\|_{L^q(\Omega)}.$$

This inequality is known as Hölder's inequality. Due to the boundedness of Ω, this inequality can be used to prove the embedding $L^p(\Omega) \rightarrow L^q(\Omega)$ for $p > q$. The inequality stays also true for the case $p = \infty$ and $q = 1$. We will mostly use the special case $p = q = 2$, in which we obtain the Cauchy-Schwarz inequality

$$\int_\Omega vw \; dx \leq \|v\|_{L^2(\Omega)} \|w\|_{L^2(\Omega)}.$$

A very important tool in the development of weak solutions for partial differential equations are Sobolev spaces. Let k be a positive integer and $p \geq 1$. We denote by $W^{k,p}(\Omega)$ the space of functions which are k-times weakly differentiable and whose derivatives lie in $L^p(\Omega)$. The Sobolev space becomes a Banach space under the norm

$$\|v\|_{W^{k,p}(\Omega)}^p := \sum_{0 \leq |\alpha| \leq k} \|D^\alpha v\|_{L^p(\Omega)}^p,$$

where $\alpha \in \mathbb{N}_0^n$ denotes a multi-index and $D^\alpha v$ its associated weak derivative. For the special case $p = 2$ we denote $H^k(\Omega) := W^{k,2}(\Omega)$. Similar to $L^2(\Omega)$ the space $H^k(\Omega)$ is a Hilbert space.

Theorem 2.1.8. *Let Ω be a bounded domain with Lipschitz boundary and $1 \leq p \leq \infty$. Then there exists a linear and continuous operator $\tau : W^{1,p}(\Omega) \rightarrow L^p(\partial\Omega)$ such that for all $y \in W^{1,p}(\Omega) \cap C(\bar{\Omega})$ it holds $(\tau y)(x) = y(x)$ almost everywhere on $\partial\Omega$.*

We are mostly interested in the case $p = 2$, which then gives the existence of an operator $\tau : H^1(\Omega) \rightarrow L^2(\partial\Omega)$. This operator τ is called trace operator. Theorem 2.1.8 can also be found in [93]. This allows us to construct the set $H_0^1(\Omega)$ of all functions in $H^1(\Omega)$ with zero boundary values

$$H_0^1(\Omega) := \{v \in H^1(\Omega) : \tau v = 0\}.$$

Sobolev functions exhibit a certain regularity, which can be characterized by the Sobolev embeddings.

Theorem 2.1.9. *Let $\Omega \subset \mathbb{R}^n$ be a bounded domain with Lipschitz boundary. Then the embedding*

$$W^{1,p}(\Omega) \rightarrow L^{p^*}(\Omega)$$

is continuous for $1 < p < n$ and $p^* = \frac{np}{n-p}$. and the embedding

$$W^{1,p}(\Omega) \rightarrow L^q(\Omega)$$

is compact for $1 \leq q < p^*$.

2.2 Ill-Posedness and Regularization Techniques

We want to solve optimization problems of the following form

$$\text{Minimize} \quad \frac{1}{2}\|Su - z\|_Y^2$$

$$\text{such that} \quad u_a \le u \le u_b \quad \text{a.e. in } \Omega. \tag{2.1}$$

Here Y is a Hilbert space and $S : L^2(\Omega) \to Y$ denotes an operator. The problem (2.1) is analysed in more detail in Section 3.1 and 3.2 for S being the solution operator of a linear partial differential equation. In Section 8.1 and 8.2 problem (2.1) is analysed for S being the solution operator of a semi-linear partial differential equation.

2.2.1 Ill-Posedness of Optimal Control Problems

We say a problem is well-posed in the sense of Hadamard [43] if a solution exists, the solution is unique and, most importantly, the solution depends continuously on the initial data. A problem is called ill-posed if it is not well-posed.

A well-posed problem is for instance the Poisson's Equation with homogeneous Dirichlet boundary condition (5.4). There, all of the three properties are satisfied.

Let X, Y be infinite dimensional Banach spaces and $S \in \mathcal{L}(X, Y)$ be compact. As S is not invertible it is not bijective. Also, in many important examples the range of S is not closed. Hence, solving the operator equation $Sx = y$ with $x \in X$ and $y \in Y$ is ill-posed because a solution do not exist in general. This motivates the auxiliary problem

$$\min_{x \in X} \frac{1}{2}\|Sx - y\|_Y^2 \tag{2.2}$$

to get as close to y as possible. The problem (2.2) is a prototypical model problem of the problems we consider in this thesis. However, this model problem is still ill-posed in general. As an example we consider the problem

$$\min_{u,y \in L^2(\Omega)} \quad J(y) = \frac{1}{2}\|y - z\|_{L^2(\Omega)}^2$$

$$\text{such that} \quad \begin{cases} -\Delta y = u & \text{in } \Omega, \\ y = 0 & \text{on } \partial\Omega. \end{cases} \tag{2.3}$$

We will treat this PDE in the next subsection and show that this is of form (2.2). It is clear, that $(\bar{u}, \bar{y})$ is a solution of (2.3) if $J(\bar{y}) = 0$, hence $\bar{y} = z$, leading to $\bar{u} = -\Delta\bar{y} = -\Delta z$. Let us now fix $\Omega = (0, \pi)$ and define for $\delta := \delta(k) := \frac{1}{k}$ with $k \in \mathbb{N}$

$$z_1(x) = 0,$$

$$z_2(x) = \delta \sin\left(\frac{x}{\delta}\right),$$

and

$$u_1(x) = -\Delta z_1(x) = 0,$$
$$u_2(x) = -\Delta z_2(x) = \frac{1}{\delta} \sin\left(\frac{x}{\delta}\right).$$

It is easy to calculate that u_1 and u_2 are solutions of (2.3) with $z = z_1$ and $z = z_2$, respectively. Note that $z_1, z_2 \in H_0^1(\Omega) \cap C^2(\bar{\Omega})$. Here z_2 can be interpreted as a sinusoidal perturbation of z_1. In fact we use this sinusoidal noise to test the stability of our algorithms, see Subsection 3.4.7 for more details. A straightforward calculation now reveals

$$\|z_1 - z_2\|_{L^2(\Omega)} \leq \pi\delta,$$
$$\|u_1 - u_2\|_{L^2(\Omega)}^2 = \frac{1}{4\delta^2}\left(2\pi - \delta \sin\left(\frac{2\pi}{\delta}\right)\right) \geq \frac{2\pi - 1}{4\delta^2}.$$

If problem (2.3) were well-posed, there would be a constant $c > 0$ such that the following inequality holds true for all $k \in \mathbb{N}$

$$\frac{2\pi - 1}{4\delta^2} \leq \|u_1 - u_2\|_{L^2(\Omega)}^2 \leq c\|z_1 - z_2\|_{L^2(\Omega)}^2 \leq c\pi^2\delta^2.$$

However, this is a contradiction for k big enough. This means that the solution does not depend continuously on the initial data. Small perturbations in the initial data of order δ lead to a perturbation of order δ^{-1} in the solution.

This violates the definition of well-posedness in the sense of Hadamard, and shows that the problem is ill-posed. This may lead to severe instabilities during the numerical calculation of a solution, as a discretization introduces small perturbations. To cope with this negative influence several different regularization methods have been introduced. In the following we want to analyse the PDE used in (2.3) followed by selected regularization methods which are used in this thesis.

2.2.2 Poisson's Equation

Let us analyse the partial differential equation

$$\begin{aligned} -\Delta y &= u \quad \text{in} \quad \Omega \\ y &= 0 \quad \text{on} \quad \partial\Omega, \end{aligned} \tag{2.4}$$

used in (2.3). It is clear that this equation cannot have a classical solution $y \in C^2(\Omega) \cap C(\bar{\Omega})$ for arbitrary $u \in L^2(\Omega)$. Instead we seek a weak solution y in the space $H_0^1(\Omega)$ defined by the following variational formulation

$$\int_\Omega \nabla y \nabla \varphi \, dx = \int_\Omega u\varphi \, dx \quad \forall \varphi \in H_0^1(\Omega).$$

This formulation is formally obtained by integration by parts and is called the weak formulation of (2.4).

Theorem 2.2.1. *Let Ω be a bounded Lipschitz domain, then for every $u \in L^2(\Omega)$ problem (2.4) has a unique weak solution $y \in H_0^1(\Omega)$. Moreover, there exists a constant $c > 0$ independent from u such that*

$$\|y\|_{H_0^1(\Omega)} \leq c\|u\|_{L^2(\Omega)}.$$

A proof can be found in [93, Theorem 2.4] and is based on the Lax-Milgram-Theorem. Furthermore Theorem 2.2.1 shows that problem (2.4) is well-posed.

Let S denote the mapping, which maps every $u \in L^2(\Omega)$ to the unique solution of (2.4). We now have the following result.

Theorem 2.2.2. *The operator $S : L^2(\Omega) \rightarrow L^2(\Omega)$ is linear, continuous, injective and compact. Furthermore the range of S is not closed.*

Proof. The linearity follows directly by the definition of the weak formulation and the continuity is a result of Theorem 2.2.1. We now use the Rellich-Kondrachov-Theorem and obtain that $H_0^1(\Omega) \rightarrow L^2(\Omega)$ is compact. By definition $S : L^2(\Omega) \rightarrow H_0^1(\Omega)$, hence S is compact as a mapping from $L^2(\Omega) \rightarrow L^2(\Omega)$.

Assume $Su_1 = Su_2$ with $u_1, u_2 \in L^2(\Omega)$. By definition of S we obtain

$$\int_\Omega (u_1 - u_2)\varphi \; \mathrm{d}x = 0 \quad \forall \varphi \in H_0^1(\Omega),$$

leading to $u_1 = u_2$ using the fundamental lemma of calculus of variations [54, Lemma 3.2.3]. This shows that S is injective.

Let us now prove that the range of S is not closed. We know that $\mathcal{R}(S)$ is closed if and only if $\dim(\mathcal{R}(S)) < \infty$. It is clear that the space $\mathcal{C}_0 := \{f \in C^\infty(\bar{\Omega}) : f = 0 \text{ on } \partial\Omega\}$ has no finite dimension. Let $y \in \mathcal{C}_0$ and define $u := -\Delta y \in C^\infty(\Omega)$, than $y = Su$. Hence $\mathcal{C}_0 \subseteq \mathcal{R}(S)$ and therefore S has non-closed range. □

2.2.3 The Tikhonov Regularization

In the following let S be linear. A well-known regularization method is the Tikhonov regularization with some positive regularization parameter $\alpha > 0$. The regularization effect is introduced by an additional L^2-term which can be interpreted as control costs. The regularized version of (2.1) is given by

$$\text{Minimize} \quad \frac{1}{2}\|Su - z\|_Y^2 + \frac{\alpha}{2}\|u\|_{L^2(\Omega)}^2 \tag{2.5}$$

$$\text{such that} \quad u_a \leq u \leq u_b \quad \text{a.e. in } \Omega.$$

This regularization method is named after Andrey Tikhonov. It is well understood in regard to convergence for $\alpha \rightarrow 0$, perturbed data, and numerical approximations, see e.g., [33,93,96,97,99]. We just want to collect some important results. The following theorem is a well known result and we also present a similar result in Theorem 6.3.1 with additional state constraints. For a non-linear S we refer to Theorem 8.3.2 for an analogous result.

Theorem 2.2.3. *The problem (2.5) has an unique solution u_α, which satisfies the variational inequality*

$$(S^*(Su_\alpha - z) + \alpha u_\alpha, v - u_\alpha)_{L^2(\Omega)} \geq 0, \quad \forall v \in U_{\text{ad}}.$$

Furthermore denote $u^\dagger$ the minimum norm solution of (2.1), i.e.

$$u^\dagger = \arg\min\{\|u\|_{L^2(\Omega)} : u \text{ solves } (2.1)\}.$$

Then we have $u_\alpha \to u^\dagger$ in $L^2(\Omega)$ as $\alpha \to 0$.

However, for α tending to zero the Tikhonov regularized problem becomes increasingly badly conditioned. This is part of the next theorem.

Theorem 2.2.4. *Let u_α and u_α^δ be the solution of (2.5) with desired states z and z^δ respectively. Assume that $\|z - z^\delta\|_Y \leq \delta$ holds with some $\delta \geq 0$. Then it holds*

$$\|u_\alpha - u_\alpha^\delta\|_{L^2(\Omega)} \leq \frac{\delta}{\sqrt{\alpha}}.$$

Proof. We start by adding the necessary optimality conditions for u_α and u_α^δ and obtain

$$\left(S^*S(u_\alpha - u_\alpha^\delta) - S^*(z - z^\delta) + \alpha(u_\alpha - u_\alpha^\delta), u_\alpha^\delta - u_\alpha\right)_{L^2(\Omega)} \geq 0,$$

which yields

$$\begin{aligned}
\|y_\alpha - y_\alpha^\delta\|_Y^2 + \alpha\|u_\alpha - u_\alpha^\delta\|_{L^2(\Omega)}^2 &\leq (z^\delta - z, y_\alpha^\delta - y_\alpha)_Y \\
&\leq \|z^\delta - z\|_Y \|y_\alpha^\delta - y_\alpha\|_Y \\
&\leq \delta^2 + \frac{1}{4}\|y_\alpha^\delta - y_\alpha\|_Y^2.
\end{aligned}$$

From this, the result now follows. $\qquad\square$

Under a suitable regularity assumption on the minimum norm solution $u^\dagger$ of (2.1) one can derive noise estimates of the form

$$\|u^\dagger - u_\alpha^\delta\|_{L^2(\Omega)} \leq c\delta^s$$

with some $s > 0$ if $\alpha = \alpha(\delta)$ is chosen in a suitable way, see [97, Theorem 3.4]. If S is non-linear, e.g. S is the solution operator of a semi-linear partial differential equation, it is more complicated to derive such noise error estimates. See [91,92] for some stability results.

2.2.4 The Proximal Point Method

The Tikhonov regularization becomes more and more ill-conditioned if α becomes very small. However sometimes it is necessary to compute a solution with very small α to recover some interesting behaviour of the optimal control. An alternative approach is the proximal point method (PPM) introduced by Martinet [67] and developed by Rockafellar [82]. Given an iterate u_k, the next iterate u_{k+1} is determined by solving

$$\text{Minimize} \quad \frac{1}{2}\|Su - z\|_Y^2 + \alpha_{k+1}\|u - u_k\|_{L^2(\Omega)}^2$$

$$\text{such that} \quad u_a \leq u \leq u_b \quad \text{a.e. in } \Omega.$$

Due to the self-canceling effect of the regularization term, there is hope to obtain convergence without the requirement that the regularization parameters α_k tend to zero. However, in general PPM is not strongly convergent due to the example given by Güler [42], which exhibits weakly but not strongly converging iterates, see also [56]. An application of this method to optimal control problems is investigated in [83]. There exists strongly convergent modifications of PPM, see e.g., [79,80,88]. Here, it is an open question how to transfer these methods to our problem while exploiting its particular structure.

In the inverse problems community this method is known as iterated Tikhonov regularization [33,45]. If one assumes attainability of z, that is, z is in the range of S and in addition a so-called source condition, convergence rates can be derived. This condition is an abstract smoothness condition closely related to the existence of Lagrange multipliers of an associated minimum-norm problem, see Section 3.3 for more details and references. While the PPM and thus the iterated Tikhonov method allow to prove beautiful monotonicity properties, we were not able to show strong convergence under conditions adapted to our situation (control constraints and non-attainability).

2.2.5 The Iterative Bregman Method

In the proximal point method a Hilbert space norm was used to regularize the problem. However this Hilbert space norm can be replaced by a different regularization term. Another suitable regularization term is the Bregman distance [9], which will be introduced in detail in Subsection 3.4.1. The resulting method is called iterative Bregman regularization method and it will be analysed in detail in Chapter 3, 4 and 5.

There, the iterate u_{k+1} is given by the solution of

$$\text{Minimize} \quad \frac{1}{2}\|Su - z\|_Y^2 + \alpha_{k+1}D^{\lambda_k}(u, u_k),$$

where $D^\lambda(u,v) = J(u) - J(v) - (u - v, \lambda)$ is called the (generalized) Bregman distance associated to a regularization function J with subgradient $\lambda \in \partial J(v)$. This iteration method was used in e.g. [11,72], where it was applied to an image restoration problem with J being the total variation. Note that for the special choice $J(u) = \frac{1}{2}\|u\|^2_{L^2(\Omega)}$ the PPM algorithm is obtained.

We choose to incorporate the control constraint into the regularization functional, resulting in

$$J(u) := \frac{1}{2}\|u\|^2_{L^2(\Omega)} + I_{U_{ad}}(u),$$

where $U_{ad} = \{u \in L^2(\Omega) : u_a \leq u \leq u_b\}$, and I is the indicator function of convex analysis. The incorporation of the explicit control constraints $u \in U_{ad}$ into the regularization functional will prove advantageous for the convergence analysis.

2.2.6 The Augmented Lagrange Method

For problems with control constraints of the following form

$$\text{Minimize} \quad \frac{1}{2}\|Su - z\|^2_Y + \frac{\alpha}{2}\|u\|^2_{L^2(\Omega)}$$

$$\text{such that} \quad u_a \leq u \leq u_b \quad \text{a.e. in } \Omega,$$

many powerful numerical methods exists. We only want to mention the semi-smooth Newton method [46, 47] and the active-set methods [4, 52]. However, in many application it is necessary to add additional state constraints. Hence the problem is of the form

$$\text{Minimize} \quad \frac{1}{2}\|Su - z\|^2_Y + \frac{\alpha}{2}\|u\|^2_{L^2(\Omega)}$$

$$\text{such that} \quad u_a \leq u \leq u_b \quad \text{a.e. in } \Omega,$$

$$Su \leq \psi,$$

with a given function $\psi \in C(\bar{\Omega})$. Here, several new problems arise. First, we need an additional regularity assumption to prove the existence of Lagrange multipliers, which are needed to establish first order optimality conditions. It turns out that the Lagrange multipliers lie in the space of Borel measures. This makes it numerically very challenging.

The augmented Lagrange method (ALM) now eliminates the additional state constraints by adding a penalty term with respect to the state constraints and an approximation of the associated multiplier. Hence we solve the problem

$$\text{Minimize} \quad \frac{1}{2}\|Su - z\|^2_Y + \frac{\alpha}{2}\|u\|^2_{L^2(\Omega)} + \frac{1}{2\rho}\int_\Omega \left((\mu + \rho(Su - \psi))_+\right)^2 - \mu^2 \, dx,$$

$$\text{such that} \quad u_a \leq u \leq u_b.$$

Here $\mu \in L^2(\Omega)$ is a given approximation of the Lagrange multiplier associated with the state constraints and $\rho > 0$ is the penalization parameter. After the problem is solved, an update rule is applied to construct a new μ and a new ρ. Such Lagrange method are very popular in the literature and they are applied to many different problems. In Section 6.1 we give some more insight into the details and also offer some literature.

In Chapter 6 we apply an augmented Lagrange method to solve the ill-posed state constrained optimal control problem

$$\text{Minimize} \quad \frac{1}{2}\|Su - z\|^2_{L^2(\Omega)}$$
$$\text{such that} \quad u_a \leq u \leq u_b \quad \text{a.e. in } \Omega,$$
$$Su \leq \psi.$$

An active-set method to solve the arising subproblems is presented in Chapter 7 along with several numerical examples to support our algorithm.

Part I

Linear State Equation

CHAPTER 3

Iterative Bregman Method for Control Constraints

In this chapter, we investigate and analyse the iterative Bregman method, which was briefly introduced in Subsection 2.2.5. This method has become quite popular in the field of optimization and inverse problems, see e.g. [10, 23] and the references therein.

Our aim is to analyse this method in the context of optimal control problems and to prove convergence rates under a suitable regularity assumption. In order to prove convergence, in [11] a source condition is imposed. Moreover, the analysis there relies heavily on the attainability of the desired state z. We prove convergence and convergence rates without the attainability assumption. To do so, the existing proof techniques had to be considerably extended.

Moreover, as argued in [97] a source condition is unlikely to hold in an optimal control setting if z is not attainable, i.e., there is no feasible u such that $z = Su$. In [96, 99] a regularity assumption on the active sets is used as suitable substitution of the source condition. Here, the active set denotes the subset of Ω, where the inequality constraints are active in the solution. However this assumption implies that the control constraints are active everywhere, and situations where the control constraints are inactive on a large part of Ω are not covered. To overcome this, in [97] both approaches are combined: A source condition is used on the part of the domain, where the inequality constraints are inactive, and a structural assumption is used on the active sets. We will use this combined assumption to prove convergence rates of the Bregman iteration.

We start this chapter by formulating the linear quadratic model problem in Section 3.1 and presenting necessary and sufficient optimality conditions in Section 3.2. The regularity assumption we use to prove convergence rates is established in Section 3.3. The iterative Bregman method is then analysed in Section 3.4. First we establish the needed auxiliary results, followed by a convergence analysis in Subsection 3.4.5. The main results of this subsection are Theorem 3.4.15 and 3.4.20. Next we consider noisy data in Subsection 3.4.6, which leads to an a-priori stopping rule which is established in Subsection 3.4.7. The results of this chapter can be found in condensed form in the publications [76, 77].

3.1 The Linear-Quadratic Problem

Denote by $\Omega \subset \mathbb{R}^n$ a bounded domain. We are interested in the linear-quadratic optimal control problem

$$\text{Minimize} \quad H(u) := \frac{1}{2}\|Su - z\|_Y^2 \qquad\qquad (P)$$
$$\text{such that} \quad u_a \leq u \leq u_b \quad \text{a.e. in } \Omega.$$

Here Y denotes a Hilbert space, $z \in Y$ is the given desired state and $S : L^2(\Omega) \to Y$ is a linear and continuous operator. Here, we have in mind to choose S to be the solution operator of a linear partial differential equation. In many situations the operator S is compact and has non-closed range, which makes (P) ill-posed. The inequality constraints are prescribed on the set Ω. We assume $u_a, u_b \in L^\infty(\Omega)$.

We say that the functional H is of tracking type, since the minimization problem intends to find a control u such that Su is as close as possible to the desired state z.

The optimization problem is subject to additional control constraints. Here we only consider box constraints, i.e. the feasible controls are bounded from below by a function u_a and by u_b from above. Let us define the set of admissible controls

$$U_{\text{ad}} := \{u \in L^2(\Omega) : u_a \leq u \leq u_b\}.$$

We assume $u_a \leq u_b$, hence $U_{\text{ad}} \neq \emptyset$. The model problem (P) can be interpreted in two different ways. In an optimal control setting, the unknown u is the control and the constraints are limitations arising from the underlying physical problem, e.g., temperature restriction of a heat source. The function z is the desired state, and we search for u such that Su is as close to z as possible with respect to the norm in Y. Here, the interesting situation is, when z cannot be reached due to the presence of the control constraints (non-attainability). If (P) is interpreted as an inverse problem, the unknown u represents some data to be reconstructed from the measurement z. Here the inequality constraints reflect a-priori information of the unknown u.

3.2 Optimality Conditions

The optimal control problem (P) is the prototypical version of an optimal control problem and has been studied excessively studied in the literature, see e.g. [93]. By standard arguments one can prove that this problem has a solution.

Theorem 3.2.1. *The problem (P) has a solution. If the operator S is injective, the solution is unique.*

Let us now comment on necessary optimality conditions for the problem (P). For this purpose it is useful to introduce the adjoint state

$$p = S^*(z - Su)$$

for a given control $u \in U_{ad}$. Since the problem (P) is convex, the first order necessary conditions are also sufficient. In the following we denote with $u^\dagger$ a solution of (P). Its associated state and adjoint state are denoted with $y^\dagger = Su^\dagger$ and $p^\dagger = S^*(z - Su^\dagger)$, respectively. We then have the following result.

Theorem 3.2.2. *Let $u^\dagger$ be a solution of (P) with state $y^\dagger$ and adjoint state $p^\dagger$. Then the following variational inequality holds:*

$$(-p^\dagger, u - u^\dagger)_{L^2(\Omega)} \geq 0, \quad \forall u \in U_{ad}.$$

Furthermore, we have the relation

$$u^\dagger(x) \begin{cases} = u_a(x) & \text{if } p^\dagger(x) < 0, \\ \in [u_a(x), u_b(x)] & \text{if } p^\dagger(x) = 0, \\ = u_b & \text{if } p^\dagger(x) > 0. \end{cases}$$

This result shows that the solution $u^\dagger$ can be uniquely determined from $p^\dagger$ if the set $\{x \in \Omega : p^\dagger(x) = 0\}$ has measure zero. Note that due to the strict convexity of the functional H with respect to Su, the optimal state $y^\dagger = Su^\dagger$ is uniquely defined. Hence the optimal adjoint state $p^\dagger = S^*(y^\dagger - z)$ is also unique. If $p^\dagger \neq 0$ almost everywhere in Ω we conclude that the problem (P) is uniquely solvable in this case.

The relation between the optimal control $u^\dagger$ and its associated adjoint state $p^\dagger$ described in Theorem 3.2.2 will be important in the next section.

3.3 Regularity Assumption

One aim of this thesis is not only to obtain convergence of certain algorithms – but also to establish convergence rates. However, to obtain regularization error estimates with respect to the control variable we need some a-priori smoothness information of the optimal control. In the following we want to introduce our regularity assumption, which we then use to prove error estimates.

3.3.1 Source Condition

A common assumption on a solution $u^\dagger$ is the following source condition, which is an abstract smoothness condition, see, e.g., [11, 24, 49, 70, 97, 99]. We say $u^\dagger$ satisfies the source condition (SC) if the following assumption holds.

Assumption SC (Source Condition). *Let $u^\dagger$ be a solution of (P). Assume that there exists an element $w \in Y$ such that $u^\dagger = P_{U_{ad}}(S^*w)$ holds.*

The source condition is equivalent to the existence of Lagrange multipliers for the problem

$$\min_{u \in U_{\mathrm{ad}}} \quad \frac{1}{2}\|u\|^2$$
$$\text{such that} \quad Su = y^\dagger, \tag{3.1}$$

where $y^\dagger$ is the uniquely defined optimal state of (P). To see this, consider the Lagrange function

$$\mathcal{L}(u, w) := \frac{1}{2}\|u\|^2 + (w, y^\dagger - Su)_{L^2(\Omega)}.$$

For every $u^\dagger$ satisfying $Su^\dagger = y^\dagger$ we obtain

$$\frac{\partial}{\partial w}\mathcal{L}(u^\dagger, w^\dagger) = y^\dagger - Su^\dagger = 0.$$

This means, the function $w^\dagger$ is a Lagrange multiplier if and only if

$$\frac{\partial}{\partial u}\mathcal{L}(u^\dagger, w^\dagger)(v - u^\dagger) \geq 0 \qquad \forall v \in U_{\mathrm{ad}}$$
$$\Longleftrightarrow \quad (u^\dagger - S^*w^\dagger, v - u^\dagger)_{L^2(\Omega)} \geq 0 \quad \forall v \in U_{\mathrm{ad}}$$
$$\Longleftrightarrow \quad u^\dagger = P_{U_{\mathrm{ad}}}(S^*w^\dagger).$$

Hence, if the optimal control $u^\dagger$ satisfies (SC) then it is a solution of (3.1). Moreover, as this optimization problem is uniquely solvable, it follows that there is at most one control satisfying (SC). Note that the existence of Lagrange multipliers is not guaranteed in general, as in many situations the operator S is compact or has non-closed range.

Assumption (SC) is quite common in the field of inverse problems, see e.g. [33]. In the context of iterative methods using Bregman distances Assumption (SC) has been used in [37–39].

Another widely used condition is the so called power-type source condition $u^\dagger = P_{U_{\mathrm{ad}}}\left((S^*S)^{\frac{\nu}{2}}w\right)$ for $\nu \in (0, 1)$. For an iterated Tikhonov method using this power-type source condition see [53].

3.3.2 Active Set Condition

If z is not attainable, i.e., $y^\dagger \neq z$, a solution $u^\dagger$ may be bang-bang, i.e., $u^\dagger$ is a linear combination of characteristic functions, hence discontinuous in general with $u^\dagger \notin H^1(\Omega)$. But in many examples the range of S^* is contained in $H^1(\Omega)$ or $C(\bar{\Omega})$. Hence, the source condition (SC) is too restrictive for bang-bang solutions.

In the optimal control of ordinary differential equations dealing with bang-bang solutions one typically assumes that the differentiable switching function $\sigma : [0, T] \to \mathbb{R}$ possess only finitely many zeros and that $|\dot{\sigma}(t)| > 0$ holds for all t where $\sigma(t) = 0$, see e.g. [36, 68, 73] and the references therein. This condition cannot be directly transferred to the control of partial differential equations. However, using a Taylor expansion it is easy to see, that this condition is equivalent to the existence of a constant $c > 0$ such that

$$\text{meas}(\{t \in [0, T] : |\sigma(t)| \leq \varepsilon\}) \leq c\varepsilon$$

holds for all $\varepsilon > 0$. In the context of optimal control of partial differential equations this transfers to

$$\text{meas}(\{x \in \Omega : |p(x)| \leq \varepsilon\}) \leq c\varepsilon,$$

where p is the adjoint state. Note that the adjoint state acts as a switching function, see Theorem 3.2.2.

We will thus resort to the following condition. We say that $u^\dagger$ satisfies the active set condition (ASC), if the following assumption holds. Let us recall the definition of $p^\dagger = S^*(z - Su^\dagger)$. In the following χ_U denotes the characteristic function of the set U.

Assumption ASC (Active-Set Condition). *Let $u^\dagger$ be a solution of (P) and assume that there exists a set $I \subseteq \Omega$, a function $w \in Y$, and positive constants κ, c such that the following holds*

1. *(source condition)* $I \supset \{x \in \Omega : p^\dagger(x) = 0\}$ *and*

$$\chi_I u^\dagger = \chi_I P_{U_{\text{ad}}}(S^* w),$$

2. *(structure of active set)* $A := \Omega \setminus I$ *and for all $\varepsilon > 0$*

$$\text{meas}\left(\{x \in A : 0 < |p^\dagger(x)| < \varepsilon\}\right) \leq c\varepsilon^\kappa,$$

3. *(regularity of solution)* $S^* w \in L^\infty(\Omega)$.

Remark 3.3.3. *Following [97, Remark 3.3], there exists at most one $u^\dagger \in U_{\text{ad}}$ satisfying Assumption (ASC). Furthermore, by [97, Remark 3.3] this has to be the minimal norm solution in U_{ad}, which is unique by [97, Lemma 2.7].*

This condition is used in [97]. It was applied for the case $\kappa = 1$, $I = \emptyset$ and $A = \Omega$ in [99]. The set I contains the set $\{x \in \Omega : p^\dagger(x) = 0\}$, which is the set of points where $u^\dagger(x)$ cannot be uniquely determined from $p^\dagger(x)$, compare to Theorem 3.2.2. On this set, we assume that $u^\dagger$ fulfills a local source condition, which implies that $u^\dagger$ has some extra regularity there. The set A contains the points, where the inequality constraints are active, since it holds by construction that $p^\dagger(x) \neq 0$ on A, which implies $u^\dagger(x) \in \{u_a(x), u_b(x)\}$.

The next theorems give some sufficient conditions for Assumption (ASC).

Theorem 3.3.4. *Let $\Omega \subset \mathbb{R}^n$, $n = 1, 2, 3$ be a bounded domain. Assume that $p \in C^1(\bar{\Omega})$ and*

$$\min_{x \in K} |\nabla p(x)| > 0 \quad \text{with} \quad K := \{x \in \bar{\Omega} : p(x) = 0\}$$

holds. Then there exists a constant $c > 0$ such that

$$\text{meas}(\{x \in \Omega : |p(x)| < \varepsilon\}) \leq c\varepsilon$$

holds for all $\varepsilon > 0$ small enough.

The proof can be found in [28, Lemma 1.3]. We will use this theorem to construct test examples such that Assumption (ASC) is satisfied with $A = \Omega$ and $\kappa = 1$. For the one-dimensional case this result can be generalized. We will start with this auxiliary result.

Lemma 3.3.5. *Let $\Omega \subset \mathbb{R}$, $\bar{x} \in \bar{\Omega}$, $m > 0$ and $f(x) = m(x - \bar{x})^n$ with $n \in \mathbb{N}$. Then there exists a constant $c > 0$ such that*

$$\text{meas}(\{x \in \Omega : |f(x)| < \varepsilon\}) \leq c\varepsilon^{1/n}$$

holds for all $\varepsilon > 0$ small enough.

Proof. Without loss of generality, we can assume $\Omega = (-1, 1)$, $m = 1$ and $\bar{x} = 0$. Let $\varepsilon > 0$ such that $\varepsilon^{1/n} < 1$. Then we can compute

$$\begin{aligned} \text{meas}(\{x \in \Omega : |f(x)| < \varepsilon\}) &= \text{meas}(\{x \in \Omega : |x^n| < \varepsilon\}) \\ &= \text{meas}((-\varepsilon^{1/n}, \varepsilon^{1/n})) \\ &= 2\varepsilon^{1/n}, \end{aligned}$$

which finishes the proof. $\qquad\square$

The next result presents a sufficient condition for Assumption (ASC) for one-dimensional problems.

Theorem 3.3.6. *Let $\Omega \subset \mathbb{R}$ be a bounded domain. Let $s \in \mathbb{N}$ and $p \in C^s(\bar{\Omega})$. Assume that p has finitely many zeros. Define the set $K := \{x \in \bar{\Omega} : p(x) = 0\}$. Furthermore assume that*

$$\min_{x \in K} |p^{(s)}(x)| > 0,$$

and there exists a point $\bar{x} \in K$ such that $|p^{(i)}(\bar{x})| = 0$ for all $i = 0, ..., s - 1$. Then there exists a constant $c > 0$ such that

$$\text{meas}(\{x \in \Omega : |p(x)| < \varepsilon\}) \leq c\varepsilon^{1/s}$$

holds for all $\varepsilon > 0$ small enough.

Proof. Let us define the following sets for $i = 0, ..., s - 1$

$$K_i := \{x \in K : |p^{(j)}(x)| = 0, \; j = 1, ..., i, \quad \text{and} \quad |p^{(i+1)}(x)| > 0\}.$$

Due to our assumptions we know that $K_{s-1} \neq \emptyset$, since $\bar{x} \in K_{s-1}$. Furthermore we obtain

$$K = \bigcup_{i=0}^{s-1} K_i.$$

Let now $y \in K_i$ for $i \in \{0, ..., s-1\}$. We then have $|p^{(i+1)}(y)| > 0$. Since $p \in C^s(\Omega)$ we obtain an $\varrho_i(y) > 0$ and $\theta_i(y) > 0$ such that $|p^{(i+1)}(x)| \geq \theta_i(y)$ for all $x \in \Omega$ such that $|x - y| \leq \varrho_i(y)$. Now define

$$\varrho_i := \min\{\varrho_i(y) : \; y \in K_i\} > 0,$$
$$\theta_i := \min\{\theta_i(y) : \; y \in K_i\} > 0.$$

Here we use that K contains only finitely many points. Now define

$$\varrho := \min\{\varrho_i : \; i = 0, ..., s-1\} > 0,$$
$$\theta := \min\{\theta_i : \; i = 0, ..., s-1\} > 0,$$

and let $\bar{\varepsilon} > 0$ such that the following implication holds true for all $\varepsilon \leq \bar{\varepsilon}$

$$x \in \{x \in \Omega : |p(x)| < \varepsilon\} \quad \Longrightarrow \quad \exists y \in K : |x - y| \leq \varrho.$$

Let us mention, that it is always possible to find such an ε, as p is continuous. In the following we always assume $\varepsilon \leq \bar{\varepsilon}$. We now want to rewrite the set $\{x \in \Omega : |p(x)| < \varepsilon\}$ with the help of the sets $B_\varrho(y) := \{x \in \Omega : |x - y| \leq \varrho\}$:

$$\{x \in \Omega : |p(x)| < \varepsilon\} \subseteq \bigcup_{y \in K} \{x \in B_\varrho(y) : |p(x)| < \varepsilon\}$$

$$= \bigcup_{i=0}^{s-1} \bigcup_{y \in K_i} \{x \in B_\varrho(y) : |p(x)| < \varepsilon\}. \qquad (3.2)$$

In the next step we want to estimate the measure of the set $\{x \in B_\varrho(y) : |p(x)| < \varepsilon\}$. Let $i \in \{0, ..., s-1\}$, $y \in K_i$ and $x \in B_\varrho(y)$. Fist notice, that due to the construction of ϱ we obtain

$$|p^{(i+1)}(\zeta)| \geq \theta > 0 \quad \forall \zeta \in [x, y].$$

Now we apply a Taylor expansion up to order $i + 1$ and use the definition of $y \in K_i$ to obtain a $\zeta \in [x, y]$ such that

$$|p(x)| = \left| \sum_{k=0}^{i} \frac{p^{(k)}(y)}{k!}(x - y)^k + \frac{1}{(i+1)!}p^{(i+1)}(\zeta)(x-y)^{i+1} \right|$$

$$= \left| \frac{1}{(i+1)!}p^{(i+1)}(\zeta)(x-y)^{i+1} \right|$$

$$\geq c\theta|x-y|^{i+1}.$$

Hence we have $c\theta|x-y|^{i+1} \leq |p(x)| < \varepsilon$ and we therefore conclude the inclusion

$$\{x \in B_\varrho(y) \, |p(x)| < \varepsilon\} \subseteq \{x \in B_\varrho(y) : c\theta|x-y|^{i+1} \leq \varepsilon\}.$$

Using Lemma 3.3.5 we now obtain

$$\text{meas}(\{x \in B_\varrho(y) : \ |p(x)| < \varepsilon\}) \leq \text{meas}(\{x \in B_\varrho(y) : c\theta|x-y|^{i+1} < \varepsilon\})$$
$$\leq c\varepsilon^{1/(i+1)}. \tag{3.3}$$

For ε small enough we obtain $\varepsilon^{1/j} \leq \varepsilon^{1/s}$ for all $1 \leq j \leq s$. Hence using (3.3) in (3.2) and using that $K_{s-1} \neq \emptyset$ yield the result

$$\text{meas}(\{x \in \Omega : \ |p(x)| < \varepsilon\}) \leq c\sum_{i=0}^{s-1} \varepsilon^{1/(i+1)} \leq c\varepsilon^{1/s}.$$

$\square$

Theorem 3.3.6 will be used in Section 5.2 to construct several test examples such that Assumption (ASC) is satisfied with different κ. If we set $s = 1$ in the theorem above, we obtain an one-dimensional version of Theorem 3.3.4.

3.4 The Bregman Iteration

In this section we first want to introduce the Bregman distance which is then used to formulate our iterative method. Let $J : L^2(\Omega) \to \mathbb{R}$ be a convex function. Let $u, v \in L^2(\Omega)$ and $\lambda \in \partial J(v)$. We now define the Bregman distance

$$D^\lambda(u, v) := J(u) - J(v) - (u - v, \lambda)_{L^2(\Omega)}.$$

The function J is called regularization functional. This type of function is a specific realization of a more general concept, first established by Bregman [9] in the year 1967. The standard, formal definition was given 14 years later by Censor and Lent [21] based on the work of Bregman. The Bregman distance was introduced to solve feasibility problems, but however was then adapted to proximal-like methods for optimization problems [22, 23] and for maximal monotone operators [30–32]. We refer to [34, Section 12.9] and [10] for additional historical information and references.

Osher et al. introduced proximal-type methods for image restoration based on Bregman distances in [72] and for sparse-reconstruction with convex functionals [41, 101]. More recently Morozov's principle was applied to a proximal-type method in [38].

3.4.1 Bregman Distance

We want to apply the Bregman distance with the regularization functional

$$J(u) := \frac{1}{2}\|u\|_{L^2(\Omega)}^2 + I_{U_{\mathrm{ad}}}(u),$$

where $I_{U_{\mathrm{ad}}}$ denotes the indicator function of the set U_{ad}. The Bregman distance for J at $u, v \in L^2(\Omega)$ and $\lambda \in \partial J(v)$ is defined as

$$D^\lambda(u, v) := J(u) - J(v) - (u - v, \lambda)_{L^2(\Omega)}.$$

If v is in U_{ad} then the subdifferential $\partial J(v)$ is non-empty. In this case, for every $\lambda \in \partial J(v)$ there exists a $w \in \partial I_{U_{\mathrm{ad}}}(v)$ such that $\lambda = v + w$. Using this decomposition, a small calculation reveals

$$D^\lambda(u, v) = \frac{1}{2}\|u - v\|_{L^2(\Omega)}^2 + I_{U_{\mathrm{ad}}}(u) - I_{U_{\mathrm{ad}}}(v) - (u - v, w)_{L^2(\Omega)}.$$

Let us summarize some properties of J and D:

Lemma 3.4.1. *Let $C \subseteq L^2(\Omega)$ be non-empty, closed, and convex. The functional*

$$J : L^2(\Omega) \to \mathbb{R} \cup \{+\infty\}, \quad u \mapsto \frac{1}{2}\|u\|_{L^2(\Omega)}^2 + I_C(u)$$

is convex and nonnegative. Furthermore the Bregman distance

$$D^\lambda(u, v) := J(u) - J(v) - (u - v, \lambda)_{L^2(\Omega)}, \quad \lambda \in \partial J(v)$$

is nonnegative and convex with respect to u.

Proof. The convexity and nonnegativity of J follow directly from the definition. For a fixed $v \in L^2(\Omega)$ the Bregman distance $D^\lambda(\cdot, v)$ is a sum of a convex and an affine linear function, hence convex. The nonnegativity follows directly from the definition of the subdifferential $\lambda \in \partial J(v)$. $\qquad\square$

For this specific choice of J we can explicitly compute the subgradients. Here $\partial I_{U_{\mathrm{ad}}}(v)$ is the normal cone of U_{ad} at v, which can be characterized as

$$\partial I_{U_{\mathrm{ad}}}(v) = \left\{ w \in L^2(\Omega) : \quad w(x) \begin{cases} \leq 0 & \text{if } v(x) = u_a(x) \\ = 0 & \text{if } u_a(x) < v(x) < u_b(x) \\ \geq 0 & \text{if } v(x) = u_b(x) \end{cases} \right\}.$$

Hence, we have for the Bregman distance at $v \in U_{\mathrm{ad}}$

$$D^\lambda(u, v) = \frac{1}{2}\|u - v\|_{L^2(\Omega)}^2 + I_{U_{\mathrm{ad}}}(u)$$

$$+ \int_{\{v=u_a\}} w(u_a - u) \, \mathrm{d}x + \int_{\{v=u_b\}} w(u_b - u) \, \mathrm{d}x,$$

where we abbreviated by $\{v = u_a\}$ the set $\{x \in \Omega : v(x) = u_a(x)\}$ and similar for $\{v = u_b\}$. We see that the Bregman distance adds two parts that measures u on the sets where the control constraints are active for v. Due to the properties of $w \in \partial I_{U_{\mathrm{ad}}}(v)$ we obtain

$$\frac{1}{2}\|u - v\|^2_{L^2(\Omega)} \le D^\lambda(u, v) \quad \forall u, v \in U_{\mathrm{ad}}, \ \lambda \in \partial J(v). \tag{3.4}$$

Since the subdifferential $\partial I_{U_{\mathrm{ad}}}(v)$ is not a singleton in general, the Bregman distance depends on the choice of the subgradient $w \in \partial I_{U_{\mathrm{ad}}}(v)$. In the algorithm described below we will derive a suitable choice for the subgradients $\lambda \in \partial J(u)$ and $w \in \partial I_{U_{\mathrm{ad}}}(u)$.

3.4.2 The Iterative Bregman Method

To start our algorithm we need suitable starting values $u_0 \in U_{\mathrm{ad}}$ and $\lambda_0 \in \partial J(u_0)$. We define u_0 to be the solution of the problem

$$\min_{u \in L^2(\Omega)} J(u) = \frac{1}{2}\|u\|^2_{L^2(\Omega)} + I_{U_{\mathrm{ad}}}(u),$$

which yields $u_0 = P_{U_{\mathrm{ad}}}(0)$. Furthermore, by Theorem 2.1.5, we have $0 \in \partial J(u_0)$, so we simply set $\lambda_0 = 0$. Note that all of the following results can be extended to arbitrary $u_0 \in U_{\mathrm{ad}}$ and general subgradients $\lambda_0 \in \partial J(u_0) \cap \mathcal{R}(S^*)$. The (prototypical) Bregman iteration is now defined as follows, see also [72]:

Algorithm 3.1. *Let $u_0 = P_{U_{\mathrm{ad}}}(0) \in U_{\mathrm{ad}}$, $\lambda_0 = 0 \in \partial J(u_0)$ and $k = 1$.*

1. *Solve for u_k:*

$$\text{Minimize} \quad \frac{1}{2}\|Su - z\|^2_Y + \alpha_k D^{\lambda_{k-1}}(u, u_{k-1}). \tag{3.5}$$

2. *Choose $\lambda_k \in \partial J(u_k)$.*

3. *Set $k := k + 1$, go back to 1.*

Here $(\alpha_k)_k$ is a bounded sequence of positive real numbers. If $u^\dagger$ is a solution of (P), it satisfies $u^\dagger = P_{U_{\mathrm{ad}}}(u^\dagger - \Theta S^*(Su^\dagger - z))$ with $\Theta > 0$ arbitrary. Therefore a possible stopping criterion is given by (with $\varepsilon > 0$)

$$\left\| u_k - P_{U_{\mathrm{ad}}}(u_k - \Theta S^*(Su_k - z)) \right\| \le \varepsilon,$$

with a suitable norm $\|\cdot\|$. Let us reformulate (3.5) using the definition of $D^\lambda(u, v)$:

$$\frac{1}{2}\|Su - z\|^2_Y + \alpha_k D^{\lambda_{k-1}}(u, u_{k-1})$$

$$= \frac{1}{2}\|Su - z\|^2_Y + \alpha_k \Big(\frac{1}{2}\|u\|^2_{L^2(\Omega)} + I_{U_{\mathrm{ad}}}(u) - \frac{1}{2}\|u_{k-1}\|^2_{L^2(\Omega)}$$

$$- I_{U_{\mathrm{ad}}}(u_{k-1}) - (u, \lambda_{k-1})_{L^2(\Omega)} + (u_{k-1}, \lambda_{k-1})_{L^2(\Omega)} \Big).$$

Note that $u_{k-1} \in U_{\text{ad}}$, hence $I_{U_{\text{ad}}}(u_{k-1}) = 0$. Now, by dropping the constant terms we obtain that (3.5) is equivalent to the following problem:

$$\text{Minimize} \quad \frac{1}{2}\|Su - z\|_Y^2 + \alpha_k \left(\frac{1}{2}\|u\|_{L^2(\Omega)}^2 - (\lambda_{k-1}, u)_{L^2(\Omega)}\right) \tag{3.6}$$

$$\text{such that} \quad u \in U_{\text{ad}}.$$

In Algorithm 3.1 it remains to specify how to choose the subgradient λ_k for the next iteration. We will show that we can construct a new subgradient based on the iterates $u_1, ..., u_k$. The following result motivates the construction of the subgradient. Moreover it shows that Algorithm 3.1 is well-posed.

Lemma 3.4.2. *The problem (3.5) has a unique solution $u_k \in U_{\text{ad}}$ and there exists $w_k \in \partial I_{U_{\text{ad}}}(u_k)$ such that*

$$S^*(Su_k - z) + \alpha_k(u_k - \lambda_{k-1} + w_k) = 0,$$

or equivalently

$$u_k = P_{U_{\text{ad}}}\left(-\frac{1}{\alpha_k}S^*(Su_k - z) + \lambda_{k-1}\right).$$

Moreover, the subdifferential $\partial J(u_k)$ is non-empty.

Proof. The set of admissible functions U_{ad} is nonempty, closed, convex, and bounded, hence weakly compact. Furthermore, the function J_k defined by

$$J_k : L^2(\Omega) \to \mathbb{R}, \quad u \mapsto \frac{1}{2}\|u\|_{L^2(\Omega)}^2 - (\lambda_{k-1}, u)_{L^2(\Omega)}$$

is continuous and convex, hence it is weakly lower semi-continuous. With this definition (3.6) can be formulated as

$$\min_{u \in U_{\text{ad}}} H(u) + \alpha_k J_k(u).$$

Recall that $H(u) = \frac{1}{2}\|Su - z\|_Y^2$. Since H is convex, the function $H + \alpha_k J_k$ is convex and by the Weierstraß theorem (with respect to the weak topology) we get existence of minimizers. Since $\alpha_k > 0$ and J_k is strictly convex, minimizers are also unique. By the first-order optimality condition for (3.6) there exists $w_k \in \partial I_{U_{\text{ad}}}(u_k)$ such that

$$S^*(Su_k - z) + \alpha_k(u_k - \lambda_{k-1} + w_k) = 0.$$

The second statement follows immediately by rewriting this equality as

$$\left(-\frac{1}{\alpha_k}S^*(Su_k - z) + \lambda_{k-1} - u_k, v - u_k\right)_{L^2(\Omega)} \leq 0, \quad \forall v \in U_{\text{ad}},$$

which implies

$$u_k = P_{U_{\text{ad}}}\left(-\frac{1}{\alpha_k}S^*(Su_k - z) + \lambda_{k-1}\right).$$

Clearly, it holds $\partial J(u_k) \neq \emptyset$. $\qquad\square$

We have $\partial J(u_k) = u_k + \partial I_{U_{ad}}(u_k)$, so motivated by Lemma 3.4.2 we set

$$\lambda_k := u_k + w_k = \frac{1}{\alpha_k} S^*(z - Su_k) + \lambda_{k-1} \in \partial J(u_k). \tag{3.7}$$

An induction argument now directly yields the following result.

Lemma 3.4.3. *Let the subgradients $\lambda_k \in \partial J(u_k)$ be chosen according to (3.7). Then it holds*

$$\lambda_k = S^* \mu_k, \quad \mu_k := \sum_{i=1}^{k} \frac{1}{\alpha_i}(z - Su_i).$$

With this choice of λ_k, we see that Algorithm 3.1 can be equivalently formulated as:

Algorithm 3.2. *Let $u_0 = P_{U_{ad}}(0) \in U_{ad}$, $\mu_0 = 0$, $\lambda_0 = 0 \in \partial J(u_0)$ and $k = 1$.*

1. *Solve for u_k:*

$$\text{Minimize} \quad \frac{1}{2}\|Su - z\|_Y^2 + \alpha_k D^{\lambda_{k-1}}(u, u_{k-1}). \tag{3.8}$$

2. *Set $\mu_k := \sum_{i=1}^{k} \frac{1}{\alpha_i}(z - Su_i)$ and $\lambda_k := S^* \mu_k$.*

3. *Set $k := k+1$, go back to 1.*

As argued in [11, 72], Algorithm 3.2 is equivalent to the following algorithm:

Algorithm 3.3. *Let $\mu_0 := 0$ and $k = 1$.*

1. *Solve for u_k:*

$$\text{Minimize} \quad \frac{1}{2}\|Su - z - \alpha_k \mu_{k-1}\|_Y^2 + \frac{\alpha_k}{2}\|u\|_{L^2(\Omega)}^2$$
$$\text{such that} \quad u_k \in U_{ad}.$$

2. *Set $\mu_k = \frac{1}{\alpha_k}(z - Su_k) + \mu_{k-1}$.*

3. *Set $k := k+1$, go back to 1.*

The equivalence can be seen directly by computing the first-order necessary and sufficient optimality conditions. For a solution u_k given by Algorithm 3.2 we obtain

$$\left(S^*(Su_k - z) + \alpha_k(u_k - \lambda_{k-1}), v - u_k\right)_{L^2(\Omega)} \geq 0, \quad \forall v \in U_{ad},$$

while for an iterate $\bar{u}_k$ and resulting $\bar{\mu}_k$ of Algorithm 3.3 we get

$$\left(S^*(S\bar{u}_k - z - \alpha_k \bar{\mu}_{k-1}) + \alpha_k \bar{u}_k, v - \bar{u}_k\right)_{L^2(\Omega)} \geq 0, \quad \forall v \in U_{ad}.$$

By adding both inequalities and applying an induction, we obtain

$$\|S(u_k - \bar{u}_k)\|_Y^2 + \alpha_k\|u_k - \bar{u}_k\|_{L^2(\Omega)}^2 \leq (\alpha_k S^* \mu_{k-1} - \alpha_k \lambda_{k-1}, \bar{u}_k - u_k)_{L^2(\Omega)}.$$

By definition $\lambda_{k-1} = S^* \mu_{k-1}$ and therefore both algorithms coincide.

3.4.3 A-Priori Error Estimates

We want to show error estimates in terms of $|H(u_k) - H(u^\dagger)|$, where $u^\dagger$ is a solution of (P). The following result can be proven similar to the proof presented in [72], hence the proof is omitted here.

Lemma 3.4.4. *The iterates of Algorithm 3.2 satisfy*

$$H(u_k) \leq H(u_{k-1}).$$

Similar to [72, Theorem 3.3] we can formulate a convergence result on $(H(u_k))_k$, together with an a-priori error estimate. We introduce the quantity

$$\gamma_k := \sum_{j=1}^{k} \frac{1}{\alpha_j}.$$

Since the sequence α_j is bounded we obtain

$$\lim_{k \to \infty} \gamma_k^{-1} = 0.$$

Theorem 3.4.5. *The iterates of Algorithm 3.2 satisfy*

$$|H(u_k) - H(u^\dagger)| = \mathcal{O}\left(\gamma_k^{-1}\right).$$

Furthermore we have

$$D^{\lambda_k}(u^\dagger, u_k) \leq D^{\lambda_{k-1}}(u^\dagger, u_{k-1}) \quad and \quad \sum_{i=1}^{\infty} D^{\lambda_{i-1}}(u_i, u_{i-1}) < \infty.$$

Proof. Due to the definition of the Bregman distance, the following 3-point identity [25, 72] holds

$$D^{\lambda_k}(u, u_k) - D^{\lambda_{k-1}}(u, u_{k-1}) + D^{\lambda_{k-1}}(u_k, u_{k-1}) = (u_k - u, \lambda_k - \lambda_{k-1}).$$

This implies the inequality

$$
\begin{aligned}
D^{\lambda_k}(u, u_k) - D^{\lambda_{k-1}}(u, u_{k-1}) &+ D^{\lambda_{k-1}}(u_k, u_{k-1}) \\
&= (u_k - u, \lambda_k - \lambda_{k-1}) \\
&= \frac{1}{\alpha_k}(S^*(Su_k - z), u - u_k) \\
&\leq \frac{1}{\alpha_k}(H(u) - H(u_k)).
\end{aligned}
$$

We now set $u = u^\dagger$ and obtain

$$D^{\lambda_k}(u^\dagger, u_k) \leq D^{\lambda_{k-1}}(u^\dagger, u_{k-1}).$$

Note that we use use the optimality of $u^\dagger$ and the non-negativity of the Bregman distance here. If we instead perform a summation over k we get

$$\sum_{i=1}^{k} D^{\lambda_i}(u^\dagger, u_i) + \sum_{i=1}^{k} \left[D^{\lambda_{i-1}}(u_i, u_{i-1}) + \frac{1}{\alpha_i} \left(H(u_i) - H(u^\dagger) \right) \right]$$

$$\leq \sum_{i=1}^{k} D^{\lambda_{i-1}}(u^\dagger, u_{i-1}).$$

Now we use the fact that $H(u_i) - H(u^\dagger) \geq 0$ and $H(u_i) \leq H(u_{i-1})$ to obtain

$$D^{\lambda_k}(u^\dagger, u_k) + \sum_{i=1}^{k} D^{\lambda_{i-1}}(u_i, u_{i-1}) + |H(u_k) - H(u^\dagger)| \sum_{i=1}^{k} \frac{1}{\alpha_i} \leq D^{\lambda_0}(u^\dagger, u_0) < \infty.$$

From this the results now follow immediately. $\qquad\square$

The sequence of regularization parameters $(\alpha_k)_k$ is bounded, leading to the convergence result $H(u_k) \to H(u^\dagger)$.

The monotonicity of $D^{\lambda_k}(u^\dagger, u_k)$ will play a crucial role in the subsequent analysis. Together with the lower bound (3.4) it will allow us to prove strong convergence $u_k \to u^\dagger$ in $L^2(\Omega)$ under suitable conditions.

3.4.4 Auxiliary Estimates

In the sequel, we will denote by $(u_k)_k$ the sequence of iterates provided by Algorithm 3.2. Let us start with the following result, which will be useful in the convergence analysis later on.

Lemma 3.4.6. *Let $\beta_j \geq 0$, such that $\beta_j \to 0$. We then have*

$$\lim_{k \to \infty} \gamma_k^{-1} \sum_{j=1}^{k} \alpha_j^{-1} \beta_j = 0.$$

Proof. Let $\varepsilon > 0$ be arbitrary. Since $\beta_j \to 0$ we can choose N such that $\beta_j \leq \frac{\varepsilon}{2}$ holds for all $j \geq N$. Since $\gamma_k^{-1} \to 0$ there is $M > N$ such that

$$\gamma_k^{-1} \sum_{j=1}^{N} \alpha_j^{-1} \beta_j \leq \frac{\varepsilon}{2}$$

holds for all $k \geq M$. We compute for $k \geq M$:

$$\gamma_k^{-1} \sum_{j=1}^{k} \alpha_j^{-1} \beta_j = \gamma_k^{-1} \sum_{j=1}^{N} \alpha_j^{-1} \beta_j + \gamma_k^{-1} \sum_{j=N+1}^{k} \alpha_j^{-1} \beta_j$$

$$\leq \frac{\varepsilon}{2} + \frac{\varepsilon}{2} \gamma_k^{-1} \sum_{j=N+1}^{k} \alpha_j^{-1} \leq \frac{\varepsilon}{2} + \frac{\varepsilon}{2} \gamma_k^{-1} \gamma_k = \varepsilon,$$

which is the claim. $\qquad\square$

We also need the following result.

Lemma 3.4.7. *For all $\gamma, \kappa, c_1, c_2 > 0$ and $u, v \in L^1(\Omega)$ there exists a constant $C > 0$ independent from γ, u, v such that the inequality*

$$c_1 \|u - v\|_{L^1(\Omega)} \leq \frac{c_2 \gamma}{2} \|u - v\|_{L^1(\Omega)}^{1 + \frac{1}{\kappa}} + C\gamma^{-\kappa}$$

holds.

Proof. We use Young's inequality to prove this result. Let $q, r > 0$ such that $q^{-1} + r^{-1} = 1$. Then, for every $a, b \geq 0$ and positive $c > 0$ we obtain

$$ab = (c\gamma)(a)\left(\frac{b}{c\gamma}\right) \leq c\gamma\left(\frac{a^q}{q} + \frac{b^r}{r(c\gamma)^r}\right) = \frac{c\gamma}{q}a^q + \frac{b^r}{rc^r}\gamma^{-r+1}.$$

With the choice of

$$q := 1 + \frac{1}{\kappa}, \quad r := \kappa + 1$$

and

$$a = \|u - v\|_{L^1(\Omega)}, \quad b := c_1, \quad c := \frac{qc_2}{2}$$

the result is obtained with the constant $C := b^r(rc^r)^{-1}$. $\qquad\square$

In the case that Su_k is equal to the optimal state $y^\dagger = Su^\dagger$, the algorithm gives $u_{k+1} = u_k$, which is then a solution of (P).

Lemma 3.4.8. *Let $y^\dagger$ be the optimal state of (P). If $Su_k = y^\dagger$ then it holds $u_{k+1} = u_k$, and u_k solves (P).*

Proof. Since u_{k+1} is the minimizer of

$$\frac{1}{2}\|Su - z\|_Y^2 + \alpha_{k+1}D^{\lambda_k}(u, u_k)$$

it follows

$$\frac{1}{2}\|Su_{k+1} - z\|_Y^2 + \alpha_{k+1}D^{\lambda_k}(u_{k+1}, u_k) \leq \frac{1}{2}\|Su_k - z\|_Y^2 + \alpha_{k+1}D^{\lambda_k}(u_k, u_k)$$

$$= \frac{1}{2}\|y^\dagger - z\|_Y^2.$$

Since $y^\dagger$ is the unique optimal state of (P), it follows $\|y^\dagger - z\|_Y \leq \|Su_{k+1} - z\|_Y$, and hence we obtain $0 = D^{\lambda_k}(u_{k+1}, u_k)$. We now use (3.4) and obtain

$$\frac{1}{2}\|u_{k+1} - u_k\|_{L^2(\Omega)}^2 \leq D^{\lambda_k}(u_{k+1}, u_k) = 0,$$

from which $u_{k+1} = u_k$ follows. Since $Su_k = y^\dagger$ it follows that u_k solves (P). $\qquad\square$

If the algorithm reaches a solution of (P) after a finite number of steps, we can show that this solution satisfies the projected source condition (SC). This condition is used below in Section 3.4.5 to prove strong convergence of the iterates.

Lemma 3.4.9. *Let u_k be a solution of (P) for some k. Then there exists an element $w \in Y$ such that $u_k = P_{U_{ad}}(S^*w)$ holds.*

Proof. For $k = 0$ this is true by the definition of $u_0 = P_{U_{ad}}(0) = P_{U_{ad}}(S^*(0))$. For $k \geq 1$ we obtain from the optimality condition as stated in Lemma 3.4.2 and with the definition of λ_k the representation

$$u_k = P_{U_{ad}}(\lambda_k) = P_{U_{ad}}(S^*\mu_k),$$

which is the stated result. $\qquad\square$

Let us now prove auxiliary results that exploits the choice of the subdifferential λ_k in (3.7). They will be employed in the convergence rate estimates below.

Lemma 3.4.10. *Let $u^\dagger$ be a solution of (P). Then it holds*

$$
\frac{1}{\alpha_k} D^{\lambda_k}(u^\dagger, u_k) + \frac{1}{2\alpha_k^2}\|S(u^\dagger - u_k)\|_Y^2 + \frac{1}{2}\|v_k\|_Y^2
$$

$$
\leq \frac{1}{\alpha_k}(u^\dagger, u^\dagger - u_k)_{L^2(\Omega)} + \frac{\gamma_k}{\alpha_k}(p^\dagger, u_k - u^\dagger)_{L^2(\Omega)} + \frac{1}{2}\|v_{k-1}\|_Y^2,
\tag{3.9}
$$

where v_k is defined by

$$
v_k := \sum_{i=1}^{k} \frac{1}{\alpha_i} S(u^\dagger - u_i).
\tag{3.10}
$$

Proof. First notice that $u^\dagger \in \partial J(u^\dagger)$ holds, which follows from

$$
u^\dagger = u^\dagger + 0 \in \partial\left(\frac{1}{2}\|\cdot\|^2\right)(u^\dagger) + \partial I_{U_{ad}}(u^\dagger) \subseteq \partial J(u^\dagger).
$$

As in the proof of [11, Theorem 4.1], we consider the sum of the Bregman distances

$$
\frac{1}{\alpha_k} D^{\lambda_k}(u^\dagger, u_k) + \frac{1}{\alpha_k} D^{u^\dagger}(u_k, u^\dagger) = \frac{1}{\alpha_k}(u^\dagger - \lambda_k, u^\dagger - u_k)_{L^2(\Omega)}.
$$

Using the definitions of v_k and $p^\dagger$, we obtain

$$
\frac{1}{\alpha_k}(-\lambda_k, u^\dagger - u_k)_{L^2(\Omega)} = \frac{1}{\alpha_k}\left(\sum_{j=1}^{k}\frac{1}{\alpha_j}(Su_j - z), S(u^\dagger - u_k)\right)_Y
$$

$$
= \frac{1}{\alpha_k}\left(\sum_{j=1}^{k}\frac{1}{\alpha_j}(S(u_j - u^\dagger + u^\dagger) - z), S(u^\dagger - u_k)\right)_Y
$$

$$
= (-v_k, v_k - v_{k-1})_Y + \frac{1}{\alpha_k}\sum_{j=1}^{k}\frac{1}{\alpha_j}(Su^\dagger - z, S(u^\dagger - u_k))_Y
$$

$$
= (-v_k, v_k - v_{k-1})_Y + \frac{\gamma_k}{\alpha_k}(p^\dagger, u_k - u^\dagger)_{L^2(\Omega)}.
$$

We continue with transforming the first addend on the right-hand side

$$(-v_k, v_k - v_{k-1})_Y = \frac{1}{2}\|v_{k-1}\|_Y^2 - \frac{1}{2}\|v_k\|_Y^2 - \frac{1}{2}\|v_k - v_{k-1}\|_Y^2$$
$$= \frac{1}{2}\|v_{k-1}\|_Y^2 - \frac{1}{2}\|v_k\|_Y^2 - \frac{1}{2\alpha_k^2}\|S(u^\dagger - u_k)\|_Y^2.$$

We obtain the result by using the nonnegativity of $D^{u^\dagger}(u_k, u^\dagger)$. $\qquad\square$

Estimate (3.9) will play a key role in the convergence analysis of the algorithm. The principal idea is to sum the inequality (3.9) with respect to k. Using the monotonicity of the Bregman distance $D^{\lambda_k}(u^\dagger, u_k)$ and inequality (3.4), we can then conclude convergence of the iterates if we succeed in estimating the terms involving the scalar product $(u^\dagger, u^\dagger - u_k)_{L^2(\Omega)}$. Note that due to Theorem 3.2.2 the term $(p^\dagger, u_k - u^\dagger)_{L^2(\Omega)}$ is non-positive.

3.4.5 Convergence Results

In the following we want to establish several different convergence results for Algorithm 3.2. The main result of the section are the regularization error estimates under the regularity assumption (SC) and (ASC), which can be found in Theorem 3.4.15 and Theorem 3.4.20, respectively. Let us start with a general convergence result for Algorithm 3.2.

Theorem 3.4.11. *Weak limit points of the sequence* $(u_k)_k$ *generated by Algorithm 3.2 are solutions to the problem* (P). *Furthermore the iterates satisfy*

$$\sum_{i=1}^{\infty}\|u_i - u_{i-1}\|_{L^2(\Omega)}^2 < \infty.$$

Proof. Since $L^2(\Omega)$ is a Hilbert space and U_{ad} is bounded, closed and convex, it is weakly compact and weakly closed. Hence we can deduce the existence of a subsequence $u_{k_j} \rightharpoonup u^* \in U_{\mathrm{ad}}$. Furthermore H is convex and continuous, so it is weakly lower semi-continuous. By Theorem 3.4.5 we know that the sequence $(H(u_k))_k$ is converging towards $H(u^\dagger)$, hence we obtain

$$H(u^\dagger) = \liminf_{j\to\infty} H(u_{k_j}) \geq H(u^*),$$

yielding $H(u^\dagger) = H(u^*)$, since $u^\dagger$ realizes the minimum of H in U_{ad}. So u^* is a solution to the problem (P). To prove the second part we use (3.4) and the result of Theorem 3.4.5 to show

$$\sum_{i=1}^{\infty}\frac{1}{2}\|u_i - u_{i-1}\|_{L^2(\Omega)}^2 \leq \sum_{i=1}^{\infty} D^{\lambda_{i-1}}(u_i, u_{i-1}) < \infty,$$

which ends the proof. $\qquad\square$

Remark 3.4.12. *The above result resembles properties of the iterates generated by the proximal point method. There it holds $\sum_{i=1}^{\infty} \|u_i - u_{i-1}\|_{L^2(\Omega)}^2 < \infty$, see e.g. [88].*

As argued in Section 2.1, the optimal state $y^\dagger$ of (P) is uniquely determined. This allows us to prove the strong convergence of the sequence $(Su_k)_k$.

Theorem 3.4.13. *Let the sequence $(u_k)_k$ be generated by Algorithm 3.2. Then it holds*

$$Su_k \to y^\dagger$$

in Y, where $y^\dagger$ is the uniquely determined optimal state of (P).

Proof. We first show $Su_k \rightharpoonup y^\dagger$. Let $(u_{k'})_{k'}$ be a subsequence of the sequence of iterates. Due to the boundedness of U_{ad}, this sequence is bounded, and has a weakly converging subsequence $(u_{k''})_{k''}$, $u_{k''} \rightharpoonup u^*$. By Theorem 3.4.11, the limit u^* is a solution of (P). This implies $Su^* = y^\dagger$. Note that a bounded linear operator is also weakly sequentially continuous. Hence, we proved that each subsequence of $(Su_k)_k$ contains a subsequence that weakly converges to $y^\dagger$. This shows $Su_k \rightharpoonup y^\dagger$. Due to Theorem 3.4.5 and $\gamma_k^{-1} \to 0$, we have that

$$H(u_k) = \frac{1}{2}\|Su_k - z\|_Y^2 \to \frac{1}{2}\|y^\dagger - z\|_Y^2 = H(u^\dagger)$$

for every solution $u^\dagger$ of (P). Using this and Lemma 3.4.4 we get

$$0 \le \|y^\dagger - z\|_Y^2 - \|Su_k - z\|_Y^2 = \|y^\dagger\|_Y^2 - \|Su_k\|_Y^2 - \frac{1}{2}(z, y^\dagger - Su_k)_Y \to 0.$$

Hence, for every $\varepsilon > 0$ we find a $\bar{k} \in \mathbb{N}$ such that for all $\bar{k} < k \in \mathbb{N}$ we have

$$0 \le \|y^\dagger\|_Y^2 - \|Su_k\|_Y^2 - \frac{1}{2}(z, y^\dagger - Su_k)_Y \le \frac{\varepsilon}{2},$$
$$\left|\frac{1}{2}(z, y^\dagger - Su_k)_Y\right| \le \frac{\varepsilon}{2}.$$

Note that for the last inequality we used the weak convergence $Su_k \rightharpoonup y^\dagger$. We now obtain

$$-\frac{\varepsilon}{2} \le \frac{1}{2}(z, y^\dagger - Su_k)_Y \le \|y^\dagger\|_Y^2 - \|Su_k\|_Y^2 \le \frac{\varepsilon}{2} + \frac{1}{2}(z, y^\dagger - Su_k)_Y \le \varepsilon.$$

This now implies convergence of the norms $\|Su_k\|_Y \to \|y^\dagger\|_Y$. Since Y is a Hilbert space, the strong convergence $Su_k \to y^\dagger$ follows immediately. □

If we assume that the problem (P) has a unique solution $u^\dagger \in U_{ad}$ we can prove strong convergence of our algorithm.

As argued above, the solution of (P) is uniquely determined if, e.g., the operator S is injective or $p^\dagger \ne 0$ almost everywhere.

Theorem 3.4.14. *Assume that $u^\dagger \in U_{\mathrm{ad}}$ is the unique solution of (P). Then the iterates of Algorithm 3.2 satisfy*

$$\lim_{k \to \infty} \|u_k - u^\dagger\|_{L^2(\Omega)} = 0 \quad \text{and} \quad \min_{i=1,\ldots,k} \frac{1}{\alpha_i} \|S(u_i - u^\dagger)\|_Y^2 \to 0.$$

Proof. With Theorem 3.4.11 we know that each weak limit point is a solution to the problem (P). So let u^* be such a point which satisfies $H(u^\dagger) = H(u^*)$. As $u^\dagger$ is the unique solution we conclude $u^* = u^\dagger$. From every subsequence of $(u_k)_k$ we can extract a weakly converging subsequence and repeat this argumentation. Hence we can conclude weak convergence $u_k \rightharpoonup u^\dagger$ of the whole sequence.
With Lemma 3.4.10 and Theorem 3.2.2 we obtain

$$\frac{1}{2\alpha_k^2} \|S(u^\dagger - u_k)\|_Y^2 + \frac{1}{\alpha_k} D^{\lambda_k}(u^\dagger, u_k) + \frac{1}{2} \|v_k\|_Y^2 \leq \frac{1}{\alpha_k}(u^\dagger, u^\dagger - u_k)_{L^2(\Omega)} + \frac{1}{2} \|v_{k-1}\|_Y^2,$$

with v_k defined in (3.10). Summing up yields

$$\sum_{j=1}^{k} \frac{1}{2\alpha_j^2} \|S(u^\dagger - u_j)\|_Y^2 + \sum_{j=1}^{k} \frac{1}{\alpha_j} D^{\lambda_j}(u^\dagger, u_j) \leq \sum_{j=1}^{k} \alpha_j^{-1}(u^\dagger, u^\dagger - u_j)_{L^2(\Omega)},$$

where we used the convention $v_0 = 0$. We now use the monotonicity of $D^{\lambda_k}(u^\dagger, u_k)$, see Theorem 3.4.5 and the estimate $\frac{1}{2} \|u^\dagger - u_k\|_{L^2(\Omega)}^2 \leq D^{\lambda_k}(u^\dagger, u_k)$ to obtain

$$\sum_{j=1}^{k} \frac{1}{2\alpha_j^2} \|S(u^\dagger - u_j)\|_Y^2 + \frac{\gamma_k}{2} \|u^\dagger - u_k\|_Y^2 \leq \sum_{j=1}^{k} \alpha_j^{-1}(u^\dagger, u^\dagger - u_j)_{L^2(\Omega)}.$$

Multiplying this inequality with γ_k^{-1} leads to

$$\min_{j=1,\ldots,k} \frac{1}{\alpha_j} \|S(u^\dagger - u_j)\|_Y^2 + \|u^\dagger - u_k\|_{L^2(\Omega)}^2 \leq 2\gamma_k^{-1} \sum_{j=1}^{k} \alpha_j^{-1}(u^\dagger, u^\dagger - u_j)_{L^2(\Omega)}.$$

We finally obtain the result by using the weak convergence $u_k \rightharpoonup u^\dagger$ and Lemma 3.4.6. □

Under Assumption (SC) we can prove strong convergence of Algorithm 3.2 together with additional regularization error estimates.

Theorem 3.4.15. *Assume that Assumption (SC) holds for $u^\dagger$. Then the iterates of Algorithm 3.2 satisfy*

$$\|u_k - u^\dagger\|_{L^2(\Omega)}^2 = \mathcal{O}(\gamma_k^{-1}),$$

$$\min_{i=1,\ldots,k} \|S(u_i - u^\dagger)\|_Y^2 = \mathcal{O}\left(\left(\sum_{i=1}^{k} \alpha_i^{-2}\right)^{-1}\right).$$

Proof. From Lemma 3.4.10 we know

$$\frac{1}{\alpha_k}D^{\lambda_k}(u^\dagger, u_k) + \frac{1}{2\alpha_k^2}\|S(u^\dagger - u_k)\|_Y^2 + \frac{1}{2}\|v_k\|_Y^2 \leq \frac{1}{\alpha_k}(u^\dagger, u^\dagger - u_k)_{L^2(\Omega)} + \frac{1}{2}\|v_{k-1}\|_Y^2,$$

with v_k defined in (3.10). It remains to estimate $(u^\dagger, u^\dagger - u_k)_{L^2(\Omega)}$ with the help of the source condition. By the definition of the projection $u^\dagger = P_{U_{\mathrm{ad}}}(S^*w)$ we get

$$\left(u^\dagger - S^*w, v - u^\dagger\right)_{L^2(\Omega)} \geq 0 \quad \forall v \in U_{\mathrm{ad}}.$$

Since $u_k \in U_{\mathrm{ad}}$ we have

$$\frac{1}{\alpha_k}\left(u^\dagger, u^\dagger - u_k\right)_{L^2(\Omega)} \leq \frac{1}{\alpha_k}\left(S^*w, u^\dagger - u_k\right)_{L^2(\Omega)} = \frac{1}{\alpha_k}(w, S(u^\dagger - u_k))_Y$$

$$= (w, v_k - v_{k-1})_Y.$$

A small variation of Lemma 3.4.10 yields

$$\frac{1}{\alpha_k}D^{\lambda_k}(u^\dagger, u_k) + \frac{1}{2\alpha_k^2}\|S(u^\dagger - u_k)\|_Y^2 + \frac{1}{2}\|v_k - w\|_Y^2 \leq \frac{1}{2}\|v_{k-1} - w\|_Y^2.$$

Following the lines of Theorem 3.4.14 we obtain by a summation

$$\frac{1}{2}\sum_{j=1}^k \frac{1}{\alpha_j^2}\|S(u^\dagger - u_j)\|_Y^2 + \frac{\gamma_k}{2}\|u^\dagger - u_k\|_{L^2(\Omega)}^2 + \frac{1}{2}\|v_k - w\|_Y^2 \leq \frac{1}{2}\|w\|_Y^2,$$

which yields the result. □

Under the source condition (SC) we can improve Lemma 3.4.8.

Lemma 3.4.16. *Assume that $u^\dagger$ satisfies Assumption (SC). If it holds $Su_k = y^\dagger$, then it follows $u_k = u^\dagger$.*

Proof. As argued in Lemma 3.4.9, u_k fulfils (SC). Hence both u_k and $u^\dagger$ are solutions of the minimal norm problem 3.1. This problem is uniquely solvable, which yields $u_k = u^\dagger$. □

While the sequence $(\lambda_k)_k$ is unbounded in general, we can prove convergence of $\gamma_k^{-1}\lambda_k$, which is a weighted average of the sequence $(S^*(z - Su_k))_k$.

Corollary 3.4.17. *Assume that Assumption (SC) holds for $u^\dagger$. Then it holds*

$$\left\|\gamma_k^{-1}\sum_{i=1}^k \frac{1}{\alpha_i}S(u_i - u^\dagger)\right\|_Y^2 + \left\|\gamma_k^{-1}\lambda_k - p^\dagger\right\|_{L^2(\Omega)}^2 = \mathcal{O}(\gamma_k^{-2}).$$

Proof. Due to the definitions of λ_k, $p^\dagger$, and v_k defined in (3.10) it holds

$$\gamma_k^{-1}\lambda_k - p^\dagger = \gamma_k^{-1}\left(\sum_{i=1}^{k}\frac{1}{\alpha_i}S^*S(u^\dagger - u_i)\right) = \gamma_k^{-1}S^*v_k.$$

Following the lines of the proof of Theorem 3.4.15, we obtain

$$\|v_k\|_Y \leq \|v_k - w\|_Y + \|w\|_Y \leq 2\|w\|_Y,$$

which yields the claim. □

When comparing the convergence rates of Theorem 3.4.15 and Corollary 3.4.17, one sees that the norm of the weighted average $\gamma_k^{-1}\sum_{i=1}^{k}\frac{1}{\alpha_i}S(u_i - u^\dagger)$ converges faster to zero than $\min_{i=1,\ldots,k}\|S(u_i - u^\dagger)\|_Y$, since it holds $\gamma_k^2 = \left(\sum_{i=1}^{k}\alpha_i^{-1}\right)^2 > \sum_{i=1}^{k}\alpha_i^{-2}$ for $k > 1$.

In the following we will show convergence results for iterates produced by Algorithm 3.2 if we assume (ASC). The special case $I = \Omega$ is already covered by Theorem 3.4.15, since for this choice of I the Assumption (ASC) reduces to the Assumption (SC).

We now focus on the case $I \neq \Omega$, that is, if the source condition is not satisfied on the whole domain Ω.

At first, let us prove a strengthened version of the first-order optimality conditions satisfied by $u^\dagger$. We refer to [87, Lemma 1.3] for a different proof.

Lemma 3.4.18. *Let $u^\dagger$ satisfy Assumption (ASC). Then there is $c_A > 0$ such that for all $u \in U_{\mathrm{ad}}$*

$$(-p^\dagger, u - u^\dagger)_{L^2(\Omega)} \geq c_A\|u - u^\dagger\|_{L^1(A)}^{1+\frac{1}{\kappa}}$$

is satisfied.

Proof. Let $\varepsilon > 0$ be given. Let us define $A_\varepsilon := \{x \in A : |p^\dagger(x)| \geq \varepsilon\}$. Then it holds

$$-\int_\Omega p^\dagger(u - u^\dagger)\,dx \geq -\int_{A_\varepsilon} p^\dagger(u - u^\dagger)\,dx - \int_{A\setminus A_\varepsilon} p^\dagger(u - u^\dagger)\,dx$$

$$\geq \varepsilon\,\|u - u^\dagger\|_{L^1(A_\varepsilon)} - \varepsilon\,\|u - u^\dagger\|_{L^1(A\setminus A_\varepsilon)}.$$

Using Assumption (ASC) to estimate the measure of the set $A \setminus A_\varepsilon$ we proceed with

$$\varepsilon\,\|u - u^\dagger\|_{L^1(A_\varepsilon)} - \varepsilon\,\|u - u^\dagger\|_{L^1(A\setminus A_\varepsilon)}$$

$$\geq \varepsilon\,\|u - u^\dagger\|_{L^1(A)} - 2\varepsilon\,\|u - u^\dagger\|_{L^1(A\setminus A_\varepsilon)}$$

$$\geq \varepsilon\,\|u - u^\dagger\|_{L^1(A)} - 2\varepsilon\,\|u - u^\dagger\|_{L^\infty(A)}\,\mathrm{meas}(A\setminus A_\varepsilon)$$

$$\geq \varepsilon\,\|u - u^\dagger\|_{L^1(A)} - c\,\varepsilon^{\kappa+1},$$

where $c > 1$ is a constant independent of u. In the last step, we used that the control bounds are given in $L^\infty(\Omega)$. Setting $\varepsilon := c^{-2/\kappa}\|u - u^\dagger\|_{L^1(A)}^{1/\kappa}$ yields

$$(-p^\dagger, u - u^\dagger)_{L^2(\Omega)} \geq c\|u - u^\dagger\|_{L^1(A)}^{1+\frac{1}{\kappa}},$$

which is the claim. $\qquad\square$

The next step concerns the estimation of $(u^\dagger, u^\dagger - u_j)_{L^2(\Omega)}$ with the help of the source condition part of (ASC).

Lemma 3.4.19. *Let $u^\dagger$ satisfy (ASC). If $I \neq \Omega$ there exists a constant $c > 0$ such that for all $u \in U_{ad}$ it holds*

$$(u^\dagger, u^\dagger - u)_{L^2(\Omega)} \leq (S^*w, u^\dagger - u)_{L^2(\Omega)} + c\,\|u^\dagger - u\|_{L^1(A)}.$$

Proof. Since U_{ad} is defined by pointwise inequalities, the projection onto U_{ad} can be taken pointwise. This implies

$$\left(\chi_I(u^\dagger - S^*w), v - u^\dagger\right)_{L^2(\Omega)} \geq 0, \quad \forall v \in U_{ad},$$

leading to

$$(\chi_I u^\dagger, u^\dagger - u)_{L^2(\Omega)} \leq (\chi_I S^*w, u^\dagger - u)_{L^2(\Omega)}.$$

This gives

$$\begin{aligned}(u^\dagger, u^\dagger - u)_{L^2(\Omega)} &= (\chi_I u^\dagger + \chi_A u^\dagger, u^\dagger - u)_{L^2(\Omega)} \\ &\leq (\chi_I S^*w + \chi_A u^\dagger, u^\dagger - u)_{L^2(\Omega)} \\ &= \left(S^*w, \chi_I(u^\dagger - u)\right)_{L^2(\Omega)} + (\chi_A u^\dagger, u^\dagger - u)_{L^2(\Omega)}.\end{aligned}$$

Since $\chi_I = 1 - \chi_A$ we have

$$S\chi_I(u^\dagger - u) = S(1 - \chi_A)(u^\dagger - u) = S(u^\dagger - u) - S\chi_A(u^\dagger - u).$$

Hence

$$\begin{aligned}(u^\dagger, u^\dagger - u)_{L^2(\Omega)} &\leq \left(w, S(u^\dagger - u) - S\chi_A(u^\dagger - u)\right)_Y + \left(u^\dagger, \chi_A(u^\dagger - u)\right)_{L^2(\Omega)} \\ &= \left(w, S(u^\dagger - u)\right)_Y + \left(u^\dagger - S^*w, \chi_A(u^\dagger - u)\right)_{L^2(\Omega)}.\end{aligned}$$

Since on A we have $p^\dagger \neq 0$ and $u^\dagger \in L^\infty(A)$, (recall $u_a, u_b \in L^\infty(A)$) so using the regularity assumption $S^*w \in L^\infty(\Omega)$ we can estimate

$$\left(u^\dagger - S^*w, \chi_A(u^\dagger - u)\right)_{L^2(\Omega)} \leq c\|u^\dagger - u\|_{L^1(A)},$$

which is the claim. $\qquad\square$

We now have all the tools to prove strong convergence for the iterates of Algorithm 3.2.

Theorem 3.4.20. *Let $u^\dagger$ satisfy Assumption (ASC). Then the iterates of Algorithm 3.2 satisfy*

$$\|u^\dagger - u_k\|_{L^2(\Omega)}^2 = \mathcal{O}\left(\gamma_k^{-1} + \gamma_k^{-1}\sum_{j=1}^{k}\alpha_j^{-1}\gamma_j^{-\kappa}\right),$$

$$\min_{j=1,\dots,k}\|S(u^\dagger - u_j)\|_Y^2 = \mathcal{O}\left(\left(\sum_{j=1}^{k}\frac{1}{\alpha_j^2}\right)^{-1}\left(1 + \sum_{j=1}^{k}\alpha_j^{-1}\gamma_j^{-\kappa}\right)\right),$$

$$\min_{j=1,\dots,k}\|u^\dagger - u_j\|_{L^1(A)}^{1+\frac{1}{\kappa}} = \mathcal{O}\left(\left(\sum_{j=1}^{k}\frac{\gamma_j}{\alpha_j}\right)^{-1}\left(1 + \sum_{j=1}^{k}\alpha_j^{-1}\gamma_j^{-\kappa}\right)\right).$$

Proof. Using the results of Lemmas 3.4.10, 3.4.18, and 3.4.19 we obtain

$$\frac{1}{\alpha_k}D^{\lambda_k}(u^\dagger, u_k) + \frac{1}{2\alpha_k^2}\|S(u^\dagger - u_k)\|_Y^2 + \frac{1}{2}\|v_k\|_Y^2 - \frac{1}{2}\|v_{k-1}\|_Y^2$$

$$\leq \frac{1}{\alpha_k}(u^\dagger, u^\dagger - u_k)_{L^2(\Omega)} + \frac{\gamma_k}{\alpha_k}(p^\dagger, u_k - u^\dagger)_{L^2(\Omega)}$$

$$\leq \frac{1}{\alpha_k}(S^*w, u^\dagger - u_k)_{L^2(\Omega)} + \frac{c}{\alpha_k}\|u^\dagger - u_k\|_{L^1(A)} - \frac{c_A\gamma_k}{\alpha_k}\|u^\dagger - u_k\|_{L^1(A)}^{1+\frac{1}{\kappa}}$$

$$\leq (w, v_k - v_{k-1})_Y + \frac{c}{\alpha_k}\|u^\dagger - u_k\|_{L^1(A)} - \frac{c_A\gamma_k}{\alpha_k}\|u^\dagger - u_k\|_{L^1(A)}^{1+\frac{1}{\kappa}}.$$

Using Lemma 3.4.7 we find

$$\frac{c}{\alpha_k}\|u^\dagger - u_k\|_{L^1(A)} \leq \frac{c_A\gamma_k}{2\alpha_k}\|u^\dagger - u_k\|_{L^1(A)}^{1+\frac{1}{\kappa}} + c\frac{\gamma_k^{-\kappa}}{\alpha_k}.$$

This implies the estimate

$$\frac{1}{\alpha_k}D^{\lambda_k}(u^\dagger, u_k) + \frac{1}{2\alpha_k^2}\|S(u^\dagger - u_k)\|_Y^2 + \frac{c_A\gamma_k}{2\alpha_k}\|u^\dagger - u_k\|_{L^1(A)}^{1+\frac{1}{\kappa}} + \frac{1}{2}\|v_k - w\|_Y^2$$

$$\leq \frac{1}{2}\|v_{k-1} - w\|_Y^2 + c\frac{\gamma_k^{-\kappa}}{\alpha_k}.$$

Summation of this inequality together with the monotonicity of the Bregman distance gives

$$\sum_{j=1}^{k}\frac{1}{\alpha_j^2}\|S(u^\dagger - u_j)\|_Y^2 + \sum_{j=1}^{k}\frac{\gamma_j}{\alpha_j}\|u^\dagger - u_j\|_{L^1(A)}^{1+\frac{1}{\kappa}}$$

$$+ \gamma_k D^{\lambda_k}(u^\dagger, u_k) + \|v_k - w\|_Y^2 \leq c\left(1 + \sum_{j=1}^{k}\alpha_j^{-1}\gamma_j^{-\kappa}\right).$$

The claim now follows using the lower bound (3.4). □

If Assumption (ASC) is satisfied with $A = \Omega$, which implies that $u^\dagger$ has a bang-bang structure on Ω, or $w = 0$, then the estimate of Theorem 3.4.20 can be improved to

$$\|u^\dagger - u_k\|^2 \le c\, \gamma_k^{-1} \sum_{j=1}^k \alpha_j^{-1}\gamma_j^{-\kappa}.$$

Similar to Corollary 3.4.17 we can prove convergence of the weighted average $\gamma_k^{-1}\lambda_k$.

Corollary 3.4.21. *Let $u^\dagger$ satisfy (ASC). Then it holds*

$$\left\|\gamma_k^{-1}\sum_{i=1}^k \frac{1}{\alpha_i}S(u_i - u^\dagger)\right\|_Y^2 + \|\gamma_k^{-1}\lambda_k - p^\dagger\|_{L^2(\Omega)}^2 = \mathcal{O}\left(\gamma_k^{-2}\left(1 + \sum_{j=1}^k \alpha_j^{-1}\gamma_j^{-\kappa}\right)\right).$$

Proof. Following the lines of Theorem 3.4.20 we obtain

$$\|v_k\|_Y^2 \le c(\|v_k - w\|_Y^2 + \|w\|_Y^2) \le c\left(1 + \sum_{j=1}^k \alpha_j^{-1}\gamma_j^{-\kappa}\right).$$

The claim follows with the same arguments as in Corollary 3.4.17. □

Let us derive precise convergence rates, if α_k is a polynomial in k.

Corollary 3.4.22. *Let $u^\dagger$ satisfy (ASC). Suppose that α_k is given by $\alpha_k = c_\alpha k^{-s}$ with $s \ge 0$, $c_\alpha > 0$. Define*

$$s_k := \begin{cases} k^{(s+1)(1-\kappa)} & \text{if } \kappa < 1, \\ \log(k) & \text{if } \kappa = 1, \\ 1 & \text{if } \kappa > 1. \end{cases}$$

Here κ is from Assumption (ASC). Then it holds

$$\|u^\dagger - u_k\|_{L^2(\Omega)}^2 = \mathcal{O}\left(k^{-(s+1)}s_k\right),$$

$$\min_{j=1,\ldots,k} \|u^\dagger - u_j\|_{L^1(A)}^{1+\frac{1}{\kappa}} = \mathcal{O}\left(k^{-2(s+1)}s_k\right),$$

$$\min_{j=1,\ldots,k} \|S(u^\dagger - u_j)\|_Y^2 = \mathcal{O}\left(k^{-(2s+1)}s_k\right),$$

$$\|\gamma_k^{-1}\lambda_k - p^\dagger\|_{L^2(\Omega)}^2 = \mathcal{O}\left(k^{-2(s+1)}s_k\right).$$

Proof. For this choice of α_k, it is easy to see that $\gamma_k^{-1} \le ck^{-(s+1)}$. Then $\alpha_j^{-1}\gamma_j^{-\kappa} \le cj^{s-(s+1)\kappa}$ which implies that $\sum_{j=1}^k \alpha_j^{-1}\gamma_j^{-\kappa} \le ck^{(s+1)(1-\kappa)}$ if $\kappa \ne 1$ and otherwise $\sum_{j=1}^k \alpha_j^{-1}\gamma_j^{-\kappa} \le c\log(k)$ if $\kappa = 1$. If $\kappa \le 1$ then the term $\sum_{j=1}^k \alpha_j^{-1}\gamma_j^{-\kappa}$ is dominating the error estimate, while for $\kappa > 1$ this term tends to zero.

This yields

$$\|u^\dagger - u_k\|^2_{L^2(\Omega)} \le c\,\gamma_k^{-1} \left(1 + \sum_{j=1}^{k} \alpha_j^{-1}\gamma_j^{-\kappa} \right) \le c\,k^{-(s+1)} s_k$$

with

$$s_k := \begin{cases} k^{(s+1)(1-\kappa)} & \text{if } \kappa < 1, \\ \log(k) & \text{if } \kappa = 1, \\ 1 & \text{if } \kappa > 1. \end{cases}$$

Using $\sum_{j=1}^{k} \alpha_j^{-1}\gamma_j \ge ck^{2(s+1)}$ and $\sum_{j=1}^{k} \alpha_j^{-2} \ge ck^{2s+1}$, we obtain the estimates

$$\min_{j=1,\dots,k} \|u^\dagger - u_j\|^{1+\frac{1}{\kappa}}_{L^1(A)} \le c \left(\sum_{j=1}^{k} \alpha_j^{-1}\gamma_j \right)^{-1} \left(1 + \sum_{j=1}^{k} \alpha_j^{-1}\gamma_j^{-\kappa} \right)$$

$$\le ck^{-2(s+1)} s_k$$

and

$$\min_{j=1,\dots,k} \|S(u^\dagger - u_j)\|^2_Y \le c \left(\sum_{j=1}^{k} \alpha_j^{-2} \right)^{-1} \left(1 + \sum_{j=1}^{k} \alpha_j^{-1}\gamma_j^{-\kappa} \right)$$

$$\le ck^{-(2s+1)} s_k.$$

Similar we obtain with Corollary 3.4.21

$$\|\gamma_k^{-1}\lambda_k - p^\dagger\|^2_{L^2(\Omega)} \le c\gamma_k^{-2} \left(1 + \sum_{j=1}^{k} \alpha_j^{-1}\gamma_j^{-\kappa} \right)$$

$$\le ck^{-2(s+1)} s_k.$$

This finishes the proof. □

After we have established the convergence proof we can now explain why the iterative Bregman method is well suited for our regularity condition and why we do not work with the proximal point method. Following the proof of Lemma 3.4.10 we have the estimate

$$\frac{1}{2}\|u^\dagger - u_k\|^2_{L^2(\Omega)} \le D^{\lambda_k}(u^\dagger, u_k) \le (u^\dagger - \lambda_k, u^\dagger - u_k)_{L^2(\Omega)}.$$

Lemma 3.4.10 now exploits the special structure of the subgradient λ_k to decompose the term $(-\lambda_k, u^\dagger - u_k)_{L^2(\Omega)}$ with the help of the auxiliary variables v_k.

The proof of the convergence rates then mainly consists of estimating the term $(u^\dagger, u^\dagger - u_k)_{L^2(\Omega)}$ with the help of our regularity assumption. This is possible since this term resembles a projection, see Lemma 3.4.19.

Now, let $(u_k^p)_k$ denote the iterates generated by the proximal point method. If we try to copy this technique we obtain by adding the first order optimality conditions that

$$\|u^\dagger - u_k^p\|_{L^2(\Omega)}^2 \leq (u^\dagger - u_{k-1}^p, u^\dagger - u_k^p)_{L^2(\Omega)}$$

holds. From here it is not clear how to handle the term $(-u_{k-1}^p, u^\dagger - u_k^p)_{L^2(\Omega)}$ in a similar way as we did for the iterative Bregman method.

3.4.6 Noise Estimates

Recall that we want to solve an optimization problem of the following form

$$\text{Minimize} \quad \frac{1}{2}\|Su - z\|_Y^2$$

$$\text{such that} \quad u_a \leq u \leq u_b \quad \text{a.e. in } \Omega.$$

In most cases the exact desired state z will not be known exactly. Assume that only an approximation $z^\delta \approx z$ is known. This is crucial, as solutions may be unstable with respect to perturbations.

In order to overcome these difficulties, several regularization methods were developed. Let us recall the Tikhonov regularization introduced in Subsection 2.2.3. The regularized problem is given by:

$$\text{Minimize} \quad \frac{1}{2}\|Su - z^\delta\|_Y^2 + \frac{\alpha}{2}\|u\|_{L^2(\Omega)}^2$$

$$\text{such that} \quad u_a \leq u \leq u_b \quad \text{a.e. in } \Omega,$$

where z^δ with $\|z - z^\delta\|_Y \leq \delta$ is the perturbed state to the noise level $\delta \geq 0$. Here one is interested in the convergence of the solution for $(\alpha, \delta) \to 0$ under some suitable conditions. For this problem convergence results were developed in [97]. In the context of inverse problems, we refer to [33]. However, for α tending to zero, the Tikhonov regularized problem becomes increasingly ill-conditioned.

The convergence and regularization error estimates presented in Subsection 3.4.5 are formulated assuming that the value z is known exactly. If only approximations $z^\delta \approx z$ are available, there will be an accumulation of the error when Algorithm 3.2 is applied. We denote by $(u_k^\delta)_k$ the iterates generated by Algorithm 3.2 when the desired state z is replaced by z^δ. We assume that the noise level δ is known and z^δ satisfies $\|z - z^\delta\|_Y \leq \delta$. In general we cannot expect convergence of the sequence $(u_k^\delta)_k$ to a solution of the unperturbed problem due to the accumulation of the error. Our aim is to identify an optimal parameter $k(\delta)$, at which it is reasonable to stop the iteration.

Before we present our stopping rule let us recall some already known facts about the iterative Bregman method. In [37,39] the authors investigated the problem

$$\text{Minimize } J(u) \quad \text{such that} \quad Su = z.$$

The Lagrangian for this problem is given as

$$\mathcal{L}(u, p) := J(u) - \langle p, Su - z \rangle.$$

Next, they apply an augmented Lagrange method with an additional augmentation $\frac{1}{2\alpha_k}\|Su - z\|_Y^2$, which fosters the fulfillment of the constraints. The augmented Lagrange method now reads as

$$u_k := \arg\min \left(\frac{1}{2\alpha_k}\|Su - z\|_Y^2 + J(u) - \langle \mu_{k-1}, Su - z\rangle \right),$$

$$\mu_k := \mu_{k-1} + \frac{1}{\alpha_k}(z - Su_k).$$

For the choice $J(u) := \frac{1}{2}\|u\|_{L^2(\Omega)}^2 + I_{U_{\mathrm{ad}}}(u)$ the augmented Lagrange method coincides with the iterative Bregman method 3.2 described in Section 3.4.

In [39, Theorem 5.3] the authors established an a-priori stopping rule $\Gamma(\delta) \in \mathbb{N}$ such that each weak cluster point of $(u_{\Gamma(\delta)}^\delta)_{\delta \to 0}$ is a solution to the original problem. The stopping index $\Gamma(\delta)$ has to satisfy

$$\lim_{\delta \to 0} \delta^2 \sum_{j=1}^{\Gamma(\delta)} \frac{1}{\alpha_j} = 0 \quad \text{and} \quad \lim_{\delta \to 0} \sum_{j=1}^{\Gamma(\delta)} \frac{1}{\alpha_j} = \infty.$$

However, the proof relies heavily on the use of the source condition (SC) and the attainablity of z. The aim of this section is to present an a-priori stopping rule, which works under our more general assumptions.

Another widely used discrepancy principle is the Morozov's principle. This rule selects a suitable stopping index by comparing the residual $\|Su - z^\delta\|_Y$ with the known noise level δ. To be precise the stopping index is defined as

$$\Gamma_M(\delta) := \min\{n \in \mathbb{N} : \|Su_n^\delta - z^\delta\|_Y \le \rho\delta\}$$

with a fixed parameter $\rho > 0$. For an analysis of Morozov's principle applied to the iterative Bregman method we refer to [38]. However this principle is not applicable here, as it needs the residual $\|Su_n^\delta - z^\delta\|_Y$ go to zero for $\delta \to 0$, which cannot be guaranteed in general if z is not reachable.

In the following we focus on the derivation of an a-priori stopping rule. We start by establishing the following noise estimate, which will be used later to construct the stopping rule.

Lemma 3.4.23. *Let $(u_k)_k$ and $(u_k^\delta)_k$ denote the sequences generated by Algorithm 3.2 for data z and z^δ, respectively. Then it holds*

$$\sum_{i=1}^{k} \frac{1}{\alpha_i}\|u_i^\delta - u_i\|_{L^2(\Omega)}^2 \le \delta^2 \sum_{i=1}^{k} \left(\frac{1}{\alpha_i^2} + \gamma_{i-1}^2 \right).$$

Proof. We introduce the following adjoint states

$$p_k := S^*(z - Su_k), \quad p_k^\delta := S^*(z^\delta - Su_k^\delta)$$

45

for u_k and u_k^δ respectively. For the subgradients we define

$$\lambda_k := \sum_{i=1}^k \frac{1}{\alpha_i} S^*(z - Su_i), \quad \lambda_k^\delta := \sum_{i=1}^k \frac{1}{\alpha_i} S^*(z^\delta - Su_i^\delta).$$

We start by using the necessary first order optimality conditions, both for u_k and u_k^δ, see Lemma 3.4.2.

$$(-p_{k+1}^\delta + \alpha_{k+1}(u_{k+1}^\delta - \lambda_k^\delta), u_{k+1} - u_{k+1}^\delta)_{L^2(\Omega)} \geq 0,$$
$$(-p_{k+1} + \alpha_{k+1}(u_{k+1} - \lambda_k), u_{k+1}^\delta - u_{k+1})_{L^2(\Omega)} \geq 0.$$

Adding both inequalities yield

$$\alpha_{k+1}\|u_{k+1} - u_{k+1}^\delta\|_{L^2(\Omega)}^2 \leq (p_{k+1} - p_{k+1}^\delta, u_{k+1} - u_{k+1}^\delta)_{L^2(\Omega)}$$
$$+ \alpha_{k+1}(\lambda_k - \lambda_k^\delta, u_{k+1} - u_{k+1}^\delta)_{L^2(\Omega)}.$$

An estimate yields for the first term

$$(p_{k+1} - p_{k+1}^\delta, u_{k+1} - u_{k+1}^\delta)_{L^2(\Omega)} = (z - Su_{k+1} - (z^\delta - Su_{k+1}^\delta), S(u_{k+1} - u_{k+1}^\delta))_Y$$
$$= (z - z^\delta, y_{k+1} - y_{k+1}^\delta)_Y + (y_{k+1}^\delta - y_{k+1}, y_{k+1} - y_{k+1}^\delta)_Y$$
$$\leq \delta\|y_{k+1} - y_{k+1}^\delta\|_Y - \|y_{k+1} - y_{k+1}^\delta\|_Y^2,$$

while for the second term we estimate

$$(\lambda_k - \lambda_k^\delta, u_{k+1} - u_{k+1}^\delta)_{L^2(\Omega)} = \sum_{i=1}^k \frac{1}{\alpha_i}(z - z^\delta - Su_i + Su_i^\delta, y_{k+1} - y_{k+1}^\delta)_Y$$
$$= \sum_{i=1}^k \frac{1}{\alpha_i}(z - z^\delta, y_{k+1} - y_{k+1}^\delta)_Y + \sum_{i=1}^k \frac{1}{\alpha_i}(y_i^\delta - y_i, y_{k+1} - y_{k+1}^\delta)_Y$$
$$\leq \delta\gamma_k\|y_{k+1} - y_{k+1}^\delta\|_Y + \left(\sum_{i=1}^k \frac{1}{\alpha_i}(y_i^\delta - y_i), y_{k+1} - y_{k+1}^\delta\right)_Y.$$

By defining the quantity

$$v_k := \sum_{i=1}^k \frac{1}{\alpha_i}(y_i - y_i^\delta),$$

and using the equality

$$(-v_k, v_{k+1} - v_k)_Y = \frac{1}{2}\|v_k\|_Y^2 - \frac{1}{2}\|v_{k+1}\|_Y^2 + \frac{1}{2}\|v_{k+1} - v_k\|_Y^2,$$

we obtain

$$(\lambda_k - \lambda_k^\delta, u_{k+1} - u_{k+1}^\delta)_{L^2(\Omega)} \leq \delta\gamma_k\|y_{k+1} - y_{k+1}^\delta\|_Y + \alpha_{k+1}(-v_k, v_{k+1} - v_k)_Y$$
$$= \delta\gamma_k\|y_{k+1} - y_{k+1}^\delta\|_Y + \alpha_{k+1}\left(\frac{1}{2}\|v_k\|_Y^2 - \frac{1}{2}\|v_{k+1}\|_Y^2 + \frac{1}{2}\|v_{k+1} - v_k\|_Y^2\right).$$

Combining these inequalities yields

$$\alpha_{k+1}\|u_{k+1} - u_{k+1}^\delta\|_{L^2(\Omega)}^2 \leq \delta\|y_{k+1} - y_{k+1}^\delta\|_Y - \|y_{k+1} - y_{k+1}^\delta\|_Y^2$$
$$+ \alpha_{k+1}\delta\gamma_k\|y_{k+1} - y_{k+1}^\delta\|_Y$$
$$+ \alpha_{k+1}^2\left(\frac{1}{2}\|v_k\|_Y^2 - \frac{1}{2}\|v_{k+1}\|_Y^2 + \frac{1}{2}\|v_{k+1} - v_k\|_Y^2\right).$$

With

$$\|v_{k+1} - v_k\|_Y^2 = \frac{1}{\alpha_{k+1}^2}\|y_{k+1} - y_{k+1}^\delta\|_Y^2,$$

we obtain

$$\alpha_{k+1}\|u_{k+1} - u_{k+1}^\delta\|_{L^2(\Omega)}^2 \leq \delta^2 + \frac{1}{4}\|y_{k+1} - y_{k+1}^\delta\|_Y^2 - \|y_{k+1} - y_{k+1}^\delta\|_Y^2$$
$$+ \alpha_{k+1}^2\delta^2\gamma_k^2 + \frac{1}{4}\|y_{k+1} - y_{k+1}^\delta\|_Y^2$$
$$+ \alpha_{k+1}^2\left(\frac{1}{2}\|v_k\|_Y^2 - \frac{1}{2}\|v_{k+1}\|_Y^2\right) + \frac{1}{2}\|y_{k+1} - y_{k+1}^\delta\|_Y^2$$
$$= \delta^2 + \alpha_{k+1}^2\delta^2\gamma_k^2 + \alpha_{k+1}^2\left(\frac{1}{2}\|v_k\|_Y^2 - \frac{1}{2}\|v_{k+1}\|_Y^2\right).$$

Dividing everything by α_{k+1}^2 and performing a summation over k yield the result

$$\sum_{i=1}^{k}\frac{1}{\alpha_i}\|u_i^\delta - u_i\|_{L^2(\Omega)}^2 \leq \delta^2\sum_{i=1}^{k}\left(\frac{1}{\alpha_i^2} + \gamma_{i-1}^2\right).$$

□

Remark 3.4.24. *The first step of Algorithm 3.2 is precisely a Tikhonov regularization with regularization parameter α_1, so we should recover the same noise estimates. This is the case, since for $k = 1$ we obtain*

$$\|u_1^\delta - u_1\|_{L^2(\Omega)} \leq \frac{\delta}{\sqrt{\alpha_1}},$$

which is the same estimate obtained for Tikhonov with regularization parameter α_1, see Theorem 2.2.4.

Remark 3.4.25. *A slight modification of the proof above yields*

$$\frac{1}{4}\sum_{i=1}^{k}\frac{1}{\alpha_i^2}\|y_i - y_i^\delta\|_Y^2 + \sum_{i=1}^{k}\frac{1}{\alpha_i}\|u_i^\delta - u_i\|_{L^2(\Omega)}^2 \leq 2\delta^2\sum_{i=1}^{k}\left(\frac{1}{\alpha_i^2} + \gamma_{i-1}^2\right),$$

from which we recover the estimates

$$\|u_1 - u_1^\delta\|_{L^2(\Omega)} \leq c\frac{\delta}{\sqrt{\alpha_1}}, \quad \|y_1 - y_1^\delta\|_Y \leq c\delta,$$

which resembles the estimates obtained for the Tikhonov regularization but with a constant $c \geq 1$, see also [97, Theorem 3.1].

3.4.7 A-Priori Stopping Rule

We will now combine the error estimates with respect to the noise level and regularization. This will give an a-priori stopping rule with best possible convergence order. We assume that Assumption (ASC) holds for $u^\dagger$. The two estimates are given as (see Lemma 3.4.23 for the noise error and Theorem 3.4.15 for the regularization error):

$$\sum_{i=1}^{k} \frac{1}{\alpha_i} \|u_i - u_i^\delta\|_{L^2(\Omega)}^2 \leq \delta^2 \sum_{i=1}^{k} \left(\frac{1}{\alpha_i^2} + \gamma_{i-1}^2 \right),$$

$$\sum_{i=1}^{k} \frac{1}{\alpha_i} \|u^\dagger - u_i\|_{L^2(\Omega)}^2 \leq c \left(1 + \sum_{i=1}^{k} \alpha_i^{-1} \gamma_i^{-\kappa} \right).$$

We define the noise error e_k^n and the regularization error e_k^r by

$$e_k^n := \delta^2 \sum_{i=1}^{k} \left(\frac{1}{\alpha_i^2} + \gamma_{i-1}^2 \right),$$

$$e_k^r := 1 + \sum_{i=1}^{k} \alpha_i^{-1} \gamma_i^{-\kappa}.$$

In the case of Assumption (SC) we set $e_k^r := 1$. All of the following results can be easily transferred to the case that $u^\dagger$ satisfies Assumption (SC). For constant $\alpha_k = \alpha$ the noise and regularization error reduce to

$$e_k^n = \frac{\delta^2}{\alpha^2} \left(k + \frac{1}{6}(k-1)k(2k-1) \right),$$

$$e_k^r = 1 + \alpha^{\kappa-1} \sum_{i=1}^{k} i^{-\kappa}.$$

For $0 < \kappa \leq 1$ we get that $e_k^r \to \infty$. Let us motivate our stopping rule. We proceed with the next iteration as long as the noise error is below the regularization error. If this condition is violated we stop the algorithm. Hence, our stopping rule is given by finding the smallest m such that $e_m^n > \tau e_m^r$ and stop the algorithm at iteration $k(\delta) = m - 1$. Here $\tau > 0$ is a scaling factor.

The stopping rule is defined as

$$k(\delta) := \min\{k \in \mathbb{N} : e_k^n > \tau e_k^r\} - 1.$$

For the case $e_1^n > \tau e_1^r$ we obtain $k(\delta) = 0$, which reflects the case that the noise error is dominating after the first iteration. This happens only if δ is too big and we will show that $k(\delta) \neq 0$ for δ small enough. Note that $k(\delta)$ depends also on τ and $(\alpha_k)_k$, but we are suppressing the dependence due to clarity of the notation.

Lemma 3.4.26. *Let $\delta > 0$. The value $k(\delta)$ defined above is well-defined and unique. Furthermore $k(\delta) \to \infty$ as $\delta \to 0$.*

Proof. For the case $e_1^n > \tau e_1^r$, there is nothing to show. Now assume that $e_1^n \le \tau e_1^r$ holds. We now show that there exists a $\bar{k} \in \mathbb{N}$ such that $e_{\bar{k}}^n > \tau e_{\bar{k}}^r$. Assume that such a value does not exists, hence we get $e_k^n \le \tau e_k^r$ for all $k \in \mathbb{N}$. Multiplying this inequality with γ_k^{-1} yields for $k \to \infty$, see Lemma 3.4.6

$$\delta^2 \gamma_k^{-1} \sum_{i=1}^k \left(\frac{1}{\alpha_i^2} + \gamma_{i-1}^2 \right) \le \tau \gamma_k^{-1} \left(1 + \sum_{i=1}^k (\alpha_i^{-1} \gamma_i^{-\kappa}) \right) \to 0.$$

Hence the sequence $(\gamma_k^{-1} e_k^n)_k$ tends to zero. Recall that there exists a constant $C > 0$ such that $\alpha_j \le C$. Hence

$$\frac{C^2}{\alpha_j^2} \ge \frac{C}{\alpha_j} \ge 1.$$

This leads to the estimate

$$\gamma_k^{-1} \sum_{i=1}^k \frac{1}{\alpha_i^2} = C^{-1} (C\gamma_k)^{-1} \left(C^2 \sum_{i=1}^k \frac{1}{\alpha_i^2} \right)$$

$$= C^{-1} \left(\sum_{i=1}^k \frac{C}{\alpha_i} \right)^{-1} \left(\sum_{i=1}^k \frac{C^2}{\alpha_i^2} \right)$$

$$\ge C^{-1} \left(\sum_{i=1}^k \frac{C}{\alpha_i} \right)^{-1} \left(\sum_{i=1}^k \frac{C}{\alpha_i} \right)$$

$$= C^{-1}.$$

We now have a contradiction since

$$0 < \delta^2 C^{-1} \le \limsup_{k \to \infty} \delta^2 \gamma_k^{-1} \sum_{i=1}^k \frac{1}{\alpha_i^2} \le \limsup_{k \to \infty} \delta^2 \gamma_k^{-1} \sum_{i=1}^k \left(\frac{1}{\alpha_i^2} + \gamma_{i-1}^2 \right)$$

$$\le \tau \lim_{k \to \infty} \gamma_k^{-1} \left(1 + \sum_{i=1}^k (\alpha_i^{-1} \gamma_i^{-\kappa}) \right)$$

$$= 0.$$

Therefore, we know the existence of $\bar{k}$ with $e_{\bar{k}}^n > \tau e_{\bar{k}}^r$, and we can deduce the existence of a maximal $k^* < \bar{k}$ with $e_i^n \le \tau e_i^r \ \forall i \le k^*$. Setting $k(\delta) := k^*$ yields the well-posedness of $k(\delta)$.

To show the second part we assume that this is wrong, hence there exists a sequence $(\delta_n)_n$ with $\delta \to 0$ and a $\bar{k} \in \mathbb{N}$ such that $k(\delta_n) < \bar{k}$ for all $n \in \mathbb{N}$. By definition we now obtain that the following inequality holds for all $n \in \mathbb{N}$

$$\delta_n^2 \sum_{i=1}^{\bar{k}} \left(\frac{1}{\alpha_i^2} + \gamma_{i-1}^2 \right) \ge \delta_n^2 \sum_{i=1}^{k(\delta_n)+1} \left(\frac{1}{\alpha_i^2} + \gamma_{i-1}^2 \right) > \tau \left(1 + \sum_{i=1}^{k(\delta_n)+1} \alpha_i^{-1} \gamma_i^{-\kappa} \right) \ge \tau.$$

This gives a contradiction for n big enough. $\qquad\square$

If we choose $k(\delta)$ based on the principle above, we can establish the following convergence result for $u^\delta_{k(\delta)}$ as $\delta \to 0$.

Theorem 3.4.27. *Let $k(\delta)$ be given by the a-priori stopping rule presented above. If $u^\dagger$ satisfies Assumption (SC) we obtain*

$$\lim_{\delta \to 0} \|u^\dagger - u^\delta_{k(\delta)}\|_{L^2(\Omega)} = 0.$$

If $u^\dagger$ satisfies Assumption (ASC) we obtain

$$\min_{j=1,\ldots,k(\delta)} \|u^\dagger - u^\delta_j\|_{L^2(\Omega)} \to 0$$

as $\delta \to 0$.

Proof. We use the triangle inequality to obtain

$$\sum_{i=1}^{k(\delta)} \frac{1}{\alpha_i} \|u^\dagger - u^\delta_i\|^2_{L^2(\Omega)} \leq \sum_{i=1}^{k(\delta)} \frac{1}{\alpha_i} \left(\|u^\dagger - u_i\|_{L^2(\Omega)} + \|u_i - u^\delta_i\|_{L^2(\Omega)} \right)^2$$

$$\leq c \left(\sum_{i=1}^{k(\delta)} \frac{1}{\alpha_i} \|u^\dagger - u_i\|^2_{L^2(\Omega)} + \sum_{i=1}^{k(\delta)} \frac{1}{\alpha_i} \|u_i - u^\delta_i\|^2_{L^2(\Omega)} \right)$$

$$\leq c \left(e^r_{k(\delta)} + e^n_{k(\delta)} \right)$$

$$\leq c(1 + \tau) e^r_{k(\delta)}.$$

If Assumption (SC) holds, we have $e^r_{k(\delta)} = 1$ and we obtain strong convergence $u^\delta_{k(\delta)} \to u^\dagger$. If $u^\dagger$ satisfies Assumption (ASC) we cannot expect e^r_k to be bounded, so we only get the weaker result

$$\min_{i=1,\ldots,k(\delta)} \|u^\dagger - u^\delta_i\|^2_{L^2(\Omega)} \leq c(1 + \tau) \gamma^{-1}_{k(\delta)} e^r_{k(\delta)}.$$

Since $k(\delta) \to \infty$ as $\delta \to 0$ we obtain the result, see Lemma 3.4.6. $\qquad \square$

Remark 3.4.28. *We expect that strong convergence $u^\delta_{k(\delta)} \to u^\dagger$ also holds under the more general Assumption (ASC), but it is an open problem to prove it. This is supported by our numerical observations, since we observed that $u^\delta_{k(\delta)}$ is converging to $u^\dagger$ independent from τ in all our simulations. For a specific problem this can be seen in Figure 3.2.*

Remark 3.4.29. *If Assumption (SC) holds for $u^\dagger$, following the proof of Theorem 3.4.27 yield the improved result*

$$\lim_{\delta \to 0} \frac{1}{\alpha_{k(\delta)}} \|u^\dagger - u^\delta_{k(\delta)}\|^2_{L^2(\Omega)} = 0.$$

However, this is not possible for Assumption (ASC).

Let us present some numerical results. We use the implementation presented in Chapter 5 and the test problems shown in Section 5.2. We add sinusoidal noise to the desired state

$$z^\delta(x) := z(x) + \delta \sin\left(\frac{x}{\delta}\right).$$

This leads to the estimate

$$\|z - z^\delta\|_{L^2(\Omega)} \leq \delta\sqrt{\text{meas}(\Omega)}.$$

Note that we used this sinusoidal noise in Subsection 2.2.1 to demonstrate the ill-posedness of (P). Hence we expect that u_k^δ will not convergence against the exact solution $u^\dagger$. Instead we expect the typical behaviour of an ill-posed problem, i.e. the error $\|u_k^\delta - u^\dagger\|_{L^2(\Omega)}$ will first decrease until a certain value k^* followed by an increase in the error. Finding this k^* is a non-trivial task.

We compute the error $\|u_k^\delta - u^\dagger\|_{L^2(\Omega)}$ for Example 5.2.1 and Example 5.2.1. For both examples we set $h = 2 \cdot 10^{-5}$ and fix $\alpha_k = 1$. For Example 5.2.1 we set $\tau = 10^3$ and for Example 5.2.1 we set $\tau = 10$. The results can be seen in Figure 3.1.

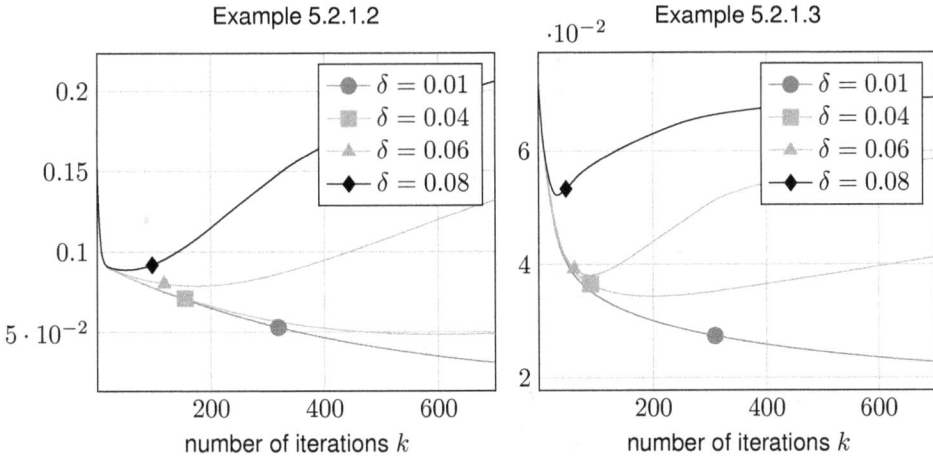

Figure 3.1: Regularization error $\|u_k^\delta - u^\dagger\|_{L^2(\Omega)}$ of Example 5.2.1 and 5.2.1, with different noise levels δ after 700 Iterations. The markers highlight the stopping points using the a-priori stopping rule.

Let us remark that such an a-priori stopping rule is barely possible in practice, as the constant κ appearing in Assumption (ASC) is not known in advance, since it depends on the unknown solution of the unregularized problem and the possible unaccessible noiseless data. Furthermore the choice of τ is not clear. If τ is chosen too large there might be some over-regularization and if τ is chosen too small some under-regularization might occur. Nevertheless we can use the a-priori rule as a benchmark to compare the convergence order of an a-posteriori stopping rule.

In Theorem 3.4.27 we proved asymptotic convergence of our stopping rule independent from τ. This can also be observed numerically. We computed $\|u_{k(\delta)}^\delta - u^\dagger\|_{L^2(\Omega)}$

for different values of τ and δ for Example 6 in Section 5.2 with constant $\alpha_k = 0.1$. The results can be found in Figure 3.2.

Furthermore we plotted the obtained values of $k(\delta)$ in Table 3.1. It is quite interesting that for fixed τ the value of $k(\frac{1}{2}\delta)$ is roughly twice the size of $k(\delta)$.

δ / τ	2^0	2^{-1}	2^{-2}	2^{-3}	2^{-4}	2^{-5}	2^{-6}	2^{-7}	2^{-8}	2^{-9}	2^{10}
10^0	1	1	1	4	6	11	19	33	60	108	194
10^1	1	3	5	9	15	27	49	88	159	288	521
10^2	4	7	13	22	40	72	130	235	426	772	1398
10^3	11	19	33	59	106	192	348	631	1143	2071	3752
10^4	27	48	87	157	285	516	934	1693	3067	5557	10067

Table 3.1: Values $k(\delta)$ for different τ and different δ for Example 6 in Section 5.2. Here we use $h = 2 \cdot 10^{-5}$ and $\alpha_k = 0.1$.

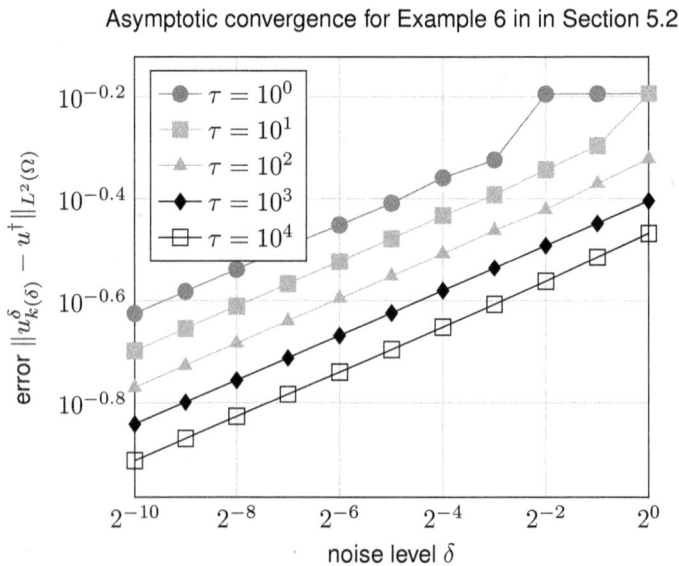

Figure 3.2: Asymptotic convergence of the stopping rule for Example 6 in in Section 5.2. Here we use $h = 2 \cdot 10^{-5}$ and $\alpha_k = 0.1$.

CHAPTER 4

The Inexact Iterative Bregman Method

In Chapter 3 the iterative Bregman method was analysed under the assumption that the arising subproblems can be solved exactly. However, in practice this will barely be possible. The aim of this chapter is to investigate our method with respect to numerical errors.

The chapter is organized as follows. In Section 4.1 we introduce a family of linear and continuous operators S_h with finite dimensional range. For $h \to 0$ we assume $S_h \to S$ in a specific way. The operator S is now replaced by the operator S_h in Algorithm 3.2. This new algorithm is then analysed in Section 4.2. In the convergence analysis we now have to take the discretization error into account, which is introduced by the operator S_h. Furthermore we allow some numerical errors in the solution of the subproblems. The main results of this chapter are Theorem 4.2.1 and 4.2.5 and the resulting corollaries. The results of this chapter can be found in the publication [75].

4.1 The Discretized Problem

The aim of this section is to introduce the operator S_h and to establish auxiliary estimates for the discretized subproblem. These estimates will then be applied to prove convergence results in section 4.2.

4.1.1 The Operator S_h

As mentioned in the introduction we want to introduce a family of linear and continuous operators $(S_h)_h$ from $L^2(\Omega)$ to Y with finite-dimensional range $Y_h \subset Y$. Throughout this chapter we make the following assumption. A similar assumption is also made in [96].

Assumption 4.1.1. *Assume that there exists a continuous and monotonically increasing function $\delta : \mathbb{R}^+ \to \mathbb{R}^+$ with $\delta(0) = 0$ such that*

$$\|(S - S_h)u_h\|_Y + \|(S^* - S_h^*)(y_h - z)\|_{L^2(\Omega)} \leq \delta(h)$$

holds for all $h \geq 0$, $u_h \in U_{\mathrm{ad}}$ and $y_h := S_h u_h$.

For the case of a linear elliptic partial differential equation, the operator S_h is the solution operator of the weak formulation with respect to the test function space Y_h. If Y_h is spanned by linear finite elements, this can be interpreted as the variational discretization in the sense of Hinze, see [46]. We consider a linear elliptic partial differential equation in Section 5.2. We assume that the operator S_h and its adjoint S_h^* can be computed exactly.

Note that Assumption 4.1.1 is an assumption on the approximation of discrete functions. Under Assumption 4.1.1 we can establish the following discretization error estimate. The proof is similar to [96, Proposition 1.6] and is omitted here.

Lemma 4.1.2. *Let u_k be the solution of*

$$\min_{u \in U_{\mathrm{ad}}} \frac{1}{2}\|Su - z\|_Y^2 + \frac{\alpha_k}{2}\|u\|_{L^2(\Omega)}^2 - \alpha_k(\lambda, u)_{L^2(\Omega)},$$

and $u_{k,h}$ be the solution of the discretized problem

$$\min_{u \in U_{\mathrm{ad}}} \frac{1}{2}\|S_h u - z\|_Y^2 + \frac{\alpha_k}{2}\|u\|_{L^2(\Omega)}^2 - \alpha_k(\lambda, u)_{L^2(\Omega)},$$

with $\lambda \in L^2(\Omega)$ and $\alpha > 0$. Then we have the following estimate

$$\frac{1}{\alpha_k}\|y_{k,h} - y_k\|_Y^2 + \|u_{k,h} - u_k\|_{L^2(\Omega)}^2 \leq c\rho_k^2\delta(h)^2$$

with the abbreviation $\rho_k^2 := \alpha_k^{-1}(1 + \alpha_k^{-1})$ and c independent from λ.

The norm of the operator S_h is bounded in the following sense.

Lemma 4.1.3. *Let $0 < h \leq h_{\max}$. Then there exists a constant $C > 0$ independent from h, such that $\|S_h\|_{L^2(\Omega) \to Y} \leq C$.*

Proof. We compute the operator norm of S_h and estimate

$$\|S_h\|_{L^2(\Omega) \to Y} = \sup_{\|u\|_{L^2(\Omega)} = 1} \|S_h u\|_Y \leq \sup_{\|u\|_{L^2(\Omega)} = 1} \left(\|(S_h - S)u\|_Y + \|Su\|_Y \right)$$

$$\leq \delta(h) + \|S\|_{L^2(\Omega) \to Y}$$

$$\leq \delta(h_{\max}) + \|S\|_{L^2(\Omega) \to Y}.$$

$\square$

In the subsequent analysis we will need the following estimate.

Lemma 4.1.4. *There exists a constant $c > 0$ independent from h, such that the following estimate holds for all $u_h \in U_{ad}$*

$$\|S_h^*(z - S_h u_h) - S^*(z - S u_h)\|_{L^2(\Omega)} \le c\delta(h).$$

Proof. We compute with $y_h := S_h u_h$

$$\|S_h^*(z - S_h u_h) - S^*(z - S u_h)\|_{L^2(\Omega)}$$
$$\le \|S_h^*(z - S_h u_h) - S^*(z - S_h u_h)\|_{L^2(\Omega)} + \|S^*(S u_h - S_h u_h)\|_{L^2(\Omega)}$$
$$\le \|(S_h^* - S^*)(z - y_h)\|_{L^2(\Omega)} + c\|(S - S_h)u_h\|_Y$$
$$\le c\delta(h).$$

Please note that we used the continuity of S^* and the assumption on the operator S_h. $\qquad\square$

An immediate consequence is the following lemma, which is needed in the subsequent analysis.

Lemma 4.1.5. *Let $u_i \in U_{ad}$ for $i = 1, .., k$. Then there exists a constant $c > 0$ independent from h and k such that the following estimate holds*

$$\left\| \sum_{i=1}^{k} \frac{1}{\alpha_i} S^*(z - S u_i) - \sum_{i=1}^{k} \frac{1}{\alpha_i} S_h^*(z - S_h u_i) \right\|_{L^2(\Omega)} \le c\gamma_k \delta(h).$$

4.1.2 A-Posteriori Error Estimate for the Discretized Subproblem

We now want to consider the discretized subproblem, i.e. we replaced the operator S in the minimization problem (step 1) of Algorithm 3.2 with the discrete operator S_h. This gives the following problem

$$\text{Minimize} \quad \frac{1}{2}\|S_h u - z\|_Y^2 + \alpha_k D^{\lambda_{k-1}}(u, u_{k-1}).$$

Similar to (3.6) this problem can be rewritten as the equivalent minimization problem (4.1). For brevity we set $\lambda := \lambda_{k-1}$ and $\alpha := \alpha_k$.

$$\text{Minimize} \quad \frac{1}{2}\|S_h u - z\|_Y^2 - \alpha(\lambda, u)_{L^2(\Omega)} + \frac{\alpha}{2}\|u\|_{L^2(\Omega)}^2, \qquad (4.1)$$
$$\text{s.t.} \quad u \in U_{ad}.$$

To construct an a-posteriori error estimate we use Theorem 2.2 in [61], which will give us the following result. Note that we also use Lemma 4.1.3 here.

Theorem 4.1.6. *Let $\hat{u}$ be the solution of the subproblem (4.1). Let $u_h \in L^2(\Omega)$ be given and define $y_h := S_h u_h$ and $p_h := S_h^*(z - y_h)$. Let $0 < h \le h_{max}$ with $h_{max} > 0$. Then there exists a constant $c > 0$ independent from h and λ such that*

$$\|u_h - \hat{u}\|_{L^2(\Omega)} \le c\left(1 + \frac{1}{\alpha}\right)\left\| u_h - P_{U_{ad}}\left(\frac{1}{\alpha}p_h + \lambda\right) \right\|_{L^2(\Omega)}.$$

This result allows us to estimate the distance to the exact solution of the subproblem. Note that the problem (4.1) is uniquely solvable if $\alpha > 0$.

For abbreviation we set

$$\mathcal{B}(\alpha, \lambda, u_h) := \left(1 + \frac{1}{\alpha}\right)\left\|u_h - P_{U_{\mathrm{ad}}}\left(\frac{1}{\alpha}S_h^*(z - S_h u_h) + \lambda\right)\right\|_{L^2(\Omega)}.$$

Let $u \in L^2(\Omega)$ be an approximate solution to the discretized subproblem (4.1). The quantity $\mathcal{B}(\alpha, \lambda, u)$ then is an upper bound for the accuracy of u. This is part of the next result. The proof follows directly with Lemma 4.1.3 and Theorem 4.1.6.

Lemma 4.1.7. *Assume that $0 < h \leq h_{\max}$. Let $\hat{u}$ be the solution of the discretized subproblem (4.1). Then there exists a constant $c > 0$ independent from h and λ such that the following implication holds for all $u \in L^2(\Omega)$ and $\varepsilon \geq 0$:*

$$\mathcal{B}(\alpha, \lambda, u) \leq \varepsilon \implies \|u - \hat{u}\|_{L^2(\Omega)} \leq c\varepsilon.$$

Let us close this section with the following remark. As mentioned in [46] the solution of the discretized subproblem (4.1) can be approximated with arbitrary accuracy. This will play a role in the analysis presented in the next section.

4.2 Inexact Bregman Iteration

Solving the subproblem

$$\text{Minimize} \quad \frac{1}{2}\|Su - z\|_Y^2 + \alpha_k D^{\lambda_{k-1}}(u, u_{k-1})$$

exactly is very costly and in general not possible. We therefore suggest the following inexact Bregman iteration which can be interpreted as an inexact version of Algorithm 3.2.

Inexact Bregman iterations are analysed in the literature, see e.g. [1,31,57,60] for a finite dimensional approach, and for an abstract Banach space setting, see [81].

Before we define the algorithm let us recall the different types of approximations we have made. The first approximation is made by the introduction of the operator S_h which reflects the discretization with respect to the mesh size h. This error can only be controlled by the reduction of the parameter h. However this will lead to a higher computational cost per iteration.

The second approximation is made during the computation of the solution of the discretized subproblem. It is not needed to compute the exact solution as we will see later in the analysis. This error is controlled by the quantity $\mathcal{B}$ and can be made arbitrarily small. We will use this fact to force convergence of the algorithm.

The inexact Bregman algorithm is now given as follows.

Algorithm 4.4. *Let* $u_0^{\text{in}} = P_{U_{\text{ad}}}(0) \in U_{\text{ad}}$, $\lambda_0^{\text{in}} = 0 \in \partial J(u_0)$ *and* $k = 1$.

1. *Find* u_k^{in} *with* $y_k^{\text{in}} = S_h u_k^{\text{in}}$ *and* $p_k^{\text{in}} = S_h^*(z - S_h u_k^{\text{in}})$ *such that*

$$\mathcal{B}(\alpha_k, \lambda_{k-1}^{\text{in}}, u_k^{\text{in}}) \leq \varepsilon_k.$$

2. *Set*

$$\lambda_k^{\text{in}} := \sum_{i=1}^{k} \frac{1}{\alpha_i} S_h^*(z - S_h u_i^{\text{in}}).$$

3. *Set* $k := k + 1$, *go back to 1.*

Here $\varepsilon_k \geq 0$ is a given sequence of positive real numbers controlling the accuracy of the approximate solution u_k^{in}. For $\varepsilon_k = 0$ for all $k \in \mathbb{N}$ and $h = 0$ Algorithm 3.2 is obtained.

The analysis of Algorithm 3.2 presented in Section 3.4 is based on the fact that $\lambda_k \in \partial J(u_k)$. This is guaranteed by the construction of λ_k. However, since $S_h \neq S$ and $\varepsilon_k > 0$ in general, we cannot expect that $\lambda_k^{\text{in}} \in \partial J(u_k^{\text{in}})$ holds.

Before we start to establish robustness results we want to give an overview over the different auxiliary problems we are going to use. Furthermore we want to introduce and clarify our notation.

4.2.1 Notation and Auxiliary Results

The aim of this subsection is to summarize the most important notations and abbreviations. Our aim is to solve the *unregularized problem*

$$\min_{u \in U_{\text{ad}}} \quad \frac{1}{2} \|Su - z\|_Y^2.$$

This problem is solvable and we want to specify a solution $u^\dagger$. We assume that this control satisfies one of the regularity assumptions (SC) or (ASC). In Algorithm 3.2 we have to solve the following *regularized problem*. We will refer to this as subproblem

$$\min_{u \in U_{\text{ad}}} \quad \frac{1}{2} \|Su - z\|_Y^2 + \frac{\alpha_{k+1}}{2} \|u\|_{L^2(\Omega)}^2 - \alpha_{k+1}(\lambda, u)_{L^2(\Omega)}, \tag{4.2}$$

with some $\lambda \in L^2(\Omega)$ and $\alpha_{k+1} > 0$. Here the (exact) unique solution is denoted with u_{k+1}^{ex}. The superscript *ex* stands for *exact solution*.

However, since the operator S is not computable in general, we introduced the operator S_h, which is an approximation of S. We now replace S with S_h in (4.2) and obtain the *discretized subproblem*

$$\min_{u \in U_{\text{ad}}} \quad \frac{1}{2} \|S_h u - z\|_Y^2 + \frac{\alpha_{k+1}}{2} \|u\|_{L^2(\Omega)}^2 - \alpha_{k+1}(\lambda, u)_{L^2(\Omega)}. \tag{4.3}$$

Again this problem is unique solvable and its solution is denoted with $u_{k+1,h}^{\text{ex}}$. The subscript h indicates that it is a *discrete* solution. Under suitable assumptions we can estimate the discretization error between u^{ex}_{k+1} and $u^{ex}_{k+1,h}$. This is done in Theorem 4.1.2.

Please note that neither u^{ex}_{k+1} nor $u^{ex}_{k+1,h}$ are computed during the algorithm. As mentioned above we can approximate $u^{ex}_{k+1,h}$ with arbitrary precision. So we compute an *inexact* solution of (4.3), which is denoted with u^{in}_{k+1}. We use the function $\mathcal{B}$ to measure the accuracy.

To control the accuracy during the algorithm we introduce a sequence $(\varepsilon_k)_k$ of positive real values. In each iteration we now search for a function $u^{in}_{k+1} \in U_{ad}$ such that $\mathcal{B}(\alpha, \lambda, u^{in}_{k+1}) \leq \varepsilon_{k+1}$.

In the end we want to estimate the error $\|u^\dagger - u^{in}_k\|_{L^2(\Omega)}$. This is done by triangular inequality

$$\|u^{in}_k - u^\dagger\|_{L^2(\Omega)} \leq \overbrace{\|u^{in}_k - u^{ex}_{k,h}\|_{L^2(\Omega)}}^{(I)} + \overbrace{\|u^{ex}_{k,h} - u^{ex}_k\|_{L^2(\Omega)}}^{(II)} + \overbrace{\|u^{ex}_k - u^\dagger\|_{L^2(\Omega)}}^{(III)}. \quad (4.4)$$

Note that (I) is controlled by the accuracy ε_k and (II) is limited by the discretization parameter h. It remains to estimate the regularization error (III) with the help of the regularity assumption (SC) or (ASC).

We also want to recall the following definitions, as they will appear quite often:

$$\gamma_k = \sum_{i=1}^{k} \frac{1}{\alpha_i}, \quad \rho_k^2 = \alpha_k^{-1}(1 + \alpha_k^{-1}).$$

4.2.2 Convergence under Source Condition

We now start to analyse Algorithm 4.4 with $u^\dagger$ satisfying Assumption (SC).

Theorem 4.2.1. *Let $u^\dagger$ satisfy Assumption (SC) and let $(\varepsilon_k)_k$ be a sequence of positive real numbers. Furthermore let $h > 0$ be given and let $(u^{in}_k)_k$ be a sequence generated by Algorithm 4.4. Then we have the estimate*

$$\sum_{i=1}^{k} \frac{1}{\alpha_i} \|u^{in}_i - u^\dagger\|^2_{L^2(\Omega)} \leq c \left(1 + \sum_{i=1}^{k} R_i + \sum_{i=1}^{k} H_i\right)$$

with the abbreviations

$$R_i := \frac{\varepsilon_i}{\alpha_i} + \frac{\varepsilon_i^2}{\alpha_i^2} + \frac{\gamma_{i-1}\varepsilon_i}{\alpha_i} + \frac{\varepsilon_i^2}{\alpha_i},$$

$$H_i := \delta(h)\left[\frac{\rho_i}{\alpha_i} + \frac{\gamma_{i-1}}{\alpha_i} + \frac{\gamma_{i-1}\rho_i}{\alpha_i}\right] + \delta(h)^2\left[\frac{\rho_i^2}{\alpha_i^2} + \frac{\rho_i^2}{\alpha_i}\right].$$

Proof. The proof is based on the splitting of the error $\|u^{in}_k - u^\dagger\|_{L^2(\Omega)}$ in three parts, see (4.4)

$$\|u^{in}_k - u^\dagger\|_{L^2(\Omega)} \leq \overbrace{\|u^{in}_k - u^{ex}_{k,h}\|_{L^2(\Omega)}}^{(I)} + \overbrace{\|u^{ex}_{k,h} - u^{ex}_k\|_{L^2(\Omega)}}^{(II)} + \overbrace{\|u^{ex}_k - u^\dagger\|_{L^2(\Omega)}}^{(III)}.$$

Here (I) is controlled by the given accuracy ε_k and (II) can be estimated with the help of Lemma 4.1.2:

$$(I) = \|u_k^{\text{in}} - u_{k,h}^{\text{ex}}\|_{L^2(\Omega)} \le c\varepsilon_k,$$
$$(II) = \|u_{k,h}^{\text{ex}} - u_k^{\text{ex}}\|_{L^2(\Omega)} \le c\rho_k\delta(h).$$

It is left to estimate (III). We start with adding the optimality conditions for u_{k+1}^{ex} and $u^\dagger$, see Lemma 3.4.2 and Theorem 3.2.2,

$$\left(S^*(Su_{k+1}^{\text{ex}} - z) + \alpha_{k+1}(u_{k+1}^{\text{ex}} - \lambda_k^{\text{in}}), v - u_{k+1}^{\text{ex}}\right)_{L^2(\Omega)} \ge 0, \quad \forall v \in U_{\text{ad}},$$
$$\left(S^*(Su^\dagger - z), v - u^\dagger\right)_{L^2(\Omega)} \ge 0, \quad \forall v \in U_{\text{ad}}.$$

Addition yields

$$\frac{1}{\alpha_{k+1}}\|S(u_{k+1}^{\text{ex}} - u^\dagger)\|_Y^2 + \|u_{k+1}^{\text{ex}} - u^\dagger\|_{L^2(\Omega)}^2 \le \left(u^\dagger - \lambda_k^{\text{in}}, u^\dagger - u_{k+1}^{\text{ex}}\right)_{L^2(\Omega)}. \tag{4.5}$$

For the term $(u^\dagger, u^\dagger - u_{k+1}^{\text{ex}})_{L^2(\Omega)}$ we estimate with help of the source condition (SC)

$$\begin{aligned}
(u^\dagger, u^\dagger - u_{k+1}^{\text{ex}})_{L^2(\Omega)} &= (u^\dagger, u^\dagger - u_{k+1}^{\text{in}})_{L^2(\Omega)} \\
&+ (u^\dagger, u_{k+1}^{\text{in}} - u_{k+1,h}^{\text{ex}})_{L^2(\Omega)} + (u^\dagger, u_{k+1,h}^{\text{ex}} - u_{k+1}^{\text{ex}})_{L^2(\Omega)} \tag{4.6} \\
&\le (S^*w, u^\dagger - u_{k+1}^{\text{in}})_{L^2(\Omega)} + c(\varepsilon_{k+1} + \rho_{k+1}\delta(h)).
\end{aligned}$$

To estimate the remaining term $(-\lambda_k^{\text{in}}, u^\dagger - u_{k+1}^{\text{ex}})_{L^2(\Omega)}$ we introduce the quantity

$$v_k^{\text{in}} := \sum_{i=1}^{k} \frac{1}{\alpha_i} S(u^\dagger - u_i^{\text{in}}).$$

This quantity will be helpful in the subsequent analysis. Let us sketch the next steps. First we will replace the operator S_h by S in order to apply the first order conditions for $u^\dagger$. Second we eliminate the unknown exact solution u_{k+1}^{ex} by its approximation u_{k+1}^{in}. For the first part we make use of Lemma 4.1.5 and estimate

$$\begin{aligned}
(-\lambda_k^{\text{in}}, u^\dagger - u_{k+1}^{\text{ex}})_{L^2(\Omega)} &= \left(\sum_{i=1}^{k} \frac{1}{\alpha_i} S_h^*(S_h u_i^{\text{in}} - z), u^\dagger - u_{k+1}^{\text{ex}}\right)_{L^2(\Omega)} \\
&= \left(\sum_{i=1}^{k} \frac{1}{\alpha_i} S^*(Su_i^{\text{in}} - z), u^\dagger - u_{k+1}^{\text{ex}}\right)_{L^2(\Omega)} \\
&+ \left(\sum_{i=1}^{k} \frac{1}{\alpha_i} S_h^*(S_h u_i^{\text{in}} - z) - \sum_{i=1}^{k} \frac{1}{\alpha_i} S^*(Su_i^{\text{in}} - z), u^\dagger - u_{k+1}^{\text{ex}}\right)_{L^2(\Omega)} \\
&\le \left(\sum_{i=1}^{k} \frac{1}{\alpha_i} S^*(Su_i^{\text{in}} - z), u^\dagger - u_{k+1}^{\text{ex}}\right)_{L^2(\Omega)} + c\gamma_k\delta(h).
\end{aligned} \tag{4.7}$$

Now we eliminate the variable z by using the first order conditions for $u^\dagger$ presented in Theorem 3.2.2.

$$\left(\sum_{i=1}^{k}\frac{1}{\alpha_i}(Su_i^{in}-z),S(u^\dagger-u_{k+1}^{ex})\right)_Y$$

$$=\sum_{i=1}^{k}\frac{1}{\alpha_i}(Su_i^{in}-Su^\dagger,S(u^\dagger-u_{k+1}^{ex}))_Y+\sum_{i=1}^{k}\frac{1}{\alpha_i}\underbrace{(Su^\dagger-z,S(u^\dagger-u_{k+1}^{ex}))_Y}_{\leq 0} \quad (4.8)$$

$$\leq\sum_{i=1}^{k}\frac{1}{\alpha_i}(S(u_i^{in}-u^\dagger),S(u^\dagger-u_{k+1}^{ex}))_Y.$$

Since the variable u_{k+1}^{ex} is unknown we replace it by its approximation u_{k+1}^{in}

$$\sum_{i=1}^{k}\frac{1}{\alpha_i}(S(u_i^{in}-u^\dagger),S(u^\dagger-u_{k+1}^{ex}))_Y$$

$$=\sum_{i=1}^{k}\frac{1}{\alpha_i}(S(u_i^{in}-u^\dagger),S(u^\dagger-u_{k+1}^{in}))_Y$$

$$+\sum_{i=1}^{k}\frac{1}{\alpha_i}(S(u_i^{in}-u^\dagger),S(u_{k+1}^{in}-u_{k+1,h}^{ex}))_Y \quad (4.9)$$

$$+\sum_{i=1}^{k}\frac{1}{\alpha_i}(S(u_i^{in}-u^\dagger),S(u_{k+1,h}^{ex}-u_{k+1}^{ex}))_Y$$

$$\leq\alpha_{k+1}(-v_k^{in},v_{k+1}^{in}-v_k^{in})_Y+c\gamma_k(\varepsilon_{k+1}+\rho_{k+1}\delta(h)).$$

Now we use (4.8) and (4.9) in (4.7) and obtain

$$(-\lambda_k^{in},u^\dagger-u_{k+1}^{ex})_Y\leq\alpha_{k+1}(-v_k^{in},v_{k+1}^{in}-v_k^{in})_Y$$

$$+c(\gamma_k\varepsilon_{k+1}+\delta(h)\gamma_k\rho_{k+1}+\delta(h)\gamma_k). \quad (4.10)$$

In the next step we plug (4.6) and (4.10) in (4.5)

$$\frac{1}{\alpha_{k+1}^2}\|S(u_{k+1}^{ex}-u^\dagger)\|_Y^2+\frac{1}{\alpha_{k+1}}\|u_{k+1}^{ex}-u^\dagger\|_{L^2(\Omega)}^2$$

$$\leq\frac{1}{\alpha_{k+1}}(u^\dagger,u^\dagger-u_{k+1}^{ex})_{L^2(\Omega)}+\frac{1}{\alpha_{k+1}}(-\lambda_k^{in},u^\dagger-u_{k+1}^{ex})_{L^2(\Omega)}$$

$$\leq\left(w,\frac{1}{\alpha_{k+1}}S(u^\dagger-u_{k+1}^{in})\right)_Y+(-v_k^{in},v_{k+1}^{in}-v_k^{in})_Y$$

$$+c\left(\frac{\varepsilon_{k+1}}{\alpha_{k+1}}+\frac{\gamma_k\varepsilon_{k+1}}{\alpha_{k+1}}\right)+c\delta(h)\left(\frac{\rho_{k+1}}{\alpha_{k+1}}+\frac{\gamma_k}{\alpha_{k+1}}+\frac{\gamma_k\rho_{k+1}}{\alpha_{k+1}}\right)$$

$$\leq(w-v_k^{in},v_{k+1}^{in}-v_k^{in})_Y$$

$$+c\left(\frac{\varepsilon_{k+1}}{\alpha_{k+1}}+\frac{\gamma_k\varepsilon_{k+1}}{\alpha_{k+1}}\right)+c\delta(h)\left(\frac{\rho_{k+1}}{\alpha_{k+1}}+\frac{\gamma_k}{\alpha_{k+1}}+\frac{\gamma_k\rho_{k+1}}{\alpha_{k+1}}\right).$$

Before we proceed we need two additional results. A calculation reveals that

$$(w - v_k^{in}, v_{k+1}^{in} - v_k^{in})_Y = \frac{1}{2}\|v_k^{in} - w\|_Y^2 - \frac{1}{2}\|v_{k+1}^{in} - w\|_Y^2 + \frac{1}{2}\|v_{k+1}^{in} - v_k^{in}\|_Y^2$$

holds. Second we obtain

$$\|S(u_{k+1}^{ex} - u_{k+1}^{in})\|_Y \le \|S(u_{k+1}^{in} - u_{k+1,h}^{ex})\|_Y + \|S(u_{k+1,h}^{ex} - u_{k+1}^{ex})\|_Y$$
$$\le c(\varepsilon_{k+1} + \rho_{k+1}\delta(h)). \tag{4.11}$$

Furthermore we use Young's inequality and (4.11) to establish for $\tau > 1$:

$$\frac{1}{2}\|v_{k+1}^{in} - v_k^{in}\|_Y^2 = \frac{1}{2\alpha_{k+1}^2}\|S(u_{k+1}^{in} - u^\dagger)\|_Y^2$$

$$\le \frac{1}{2\alpha_{k+1}^2}\left(\left(1 + \frac{1}{\tau}\right)\|S(u_{k+1}^{ex} - u^\dagger)\|_Y^2 + (1 + \tau)\|S(u_{k+1}^{ex} - u_{k+1}^{in})\|_Y^2\right)$$

$$\le \frac{1}{2\alpha_{k+1}^2}\left(1 + \frac{1}{\tau}\right)\|S(u_{k+1}^{ex} - u^\dagger)\|_Y^2 + \frac{c}{\alpha_{k+1}^2}\left(\varepsilon_{k+1}^2 + \rho_{k+1}^2\delta(h)^2\right).$$

This now yields

$$\frac{c_\tau}{\alpha_{k+1}^2}\|S(u_{k+1}^{ex} - u^\dagger)\|_Y^2 + \frac{1}{\alpha_{k+1}}\|u_{k+1}^{ex} - u^\dagger\|_{L^2(\Omega)}^2$$

$$\le \frac{1}{2}\|v_k^{in} - w\|_Y^2 - \frac{1}{2}\|v_{k+1}^{in} - w\|_Y^2 + c\left(\tilde{R}_{k+1} + \delta(h)\tilde{H}_{k+1}^{(1)} + \delta(h)^2\tilde{H}_{k+1}^{(2)}\right),$$

with $c_\tau = 1 - \frac{1}{2}\left(1 + \frac{1}{\tau}\right) > 0$ and the abbreviations

$$\tilde{R}_i := \frac{\varepsilon_i}{\alpha_i} + \frac{\varepsilon_i^2}{\alpha_i^2} + \frac{\gamma_{i-1}\varepsilon_i}{\alpha_i},$$

$$\tilde{H}_i^{(1)} := \frac{\rho_i}{\alpha_i} + \frac{\gamma_{i-1}}{\alpha_i} + \frac{\gamma_{i-1}\rho_i}{\alpha_i},$$

$$\tilde{H}_i^{(2)} := \frac{\rho_i^2}{\alpha_i^2}.$$

Summation over k finally reveals

$$c_\tau \sum_{i=1}^{k} \frac{1}{\alpha_i^2}\|S(u_i^{ex} - u^\dagger)\|_Y^2 + \sum_{i=1}^{k} \frac{1}{\alpha_i}\|u_i^{ex} - u^\dagger\|_{L^2(\Omega)}^2$$

$$\le \frac{1}{2}\|w\|_Y^2 + c\sum_{i=1}^{k}\tilde{R}_i + c\delta(h)\sum_{i=1}^{k}\tilde{H}_i^{(1)} + c\delta(h)^2\sum_{i=1}^{k}\tilde{H}_i^{(2)},$$

where we used the convention $v_0^{in} = 0$. The result now follows by triangular inequality, as mentioned in the beginning of the proof.

$$\square$$

Let us point out that the variables R_i can be identified with the accuracy of the iterates and the H_i are only influenced by the discretization. This result above can now be interpreted in different ways. First we start with the (theoretical) case that we can evaluate the operator S and its dual S^*. This refers to the case where $h = 0$.

Corollary 4.2.2. *Let $u^\dagger$ satisfy Assumption (SC) and let $(\varepsilon_k)_k$ be a sequence of positive real numbers such that*

$$\sum_{i=1}^{\infty} R_i < \infty.$$

Furthermore assume that $S_h = S$ and let $(u_k^{in})_k$ be a sequence generated by Algorithm 4.4. Then we have

$$\lim_{k \to \infty} \frac{1}{\alpha_k} \|u_k^{in} - u^\dagger\|_{L^2(\Omega)}^2 = 0.$$

This is exactly the same convergence result obtained for Algorithm 3.2, see the proof of Theorem 3.4.15. This means that the accumulation of the error introduced by solving the subproblem can be compensated by a suitable choice of the sequence $(\varepsilon_k)_k$.

The other interesting case is, that we can solve the discretized subproblem exactly, i.e. $\varepsilon_k = 0$ for all $k \in \mathbb{N}$. Here we obtain convergence in the following sense.

Corollary 4.2.3. *Let $u^\dagger$ satisfy Assumption (SC). Let $h_{\max} > 0$ be given and $\varepsilon_k = 0$ for all $k \in \mathbb{N}$. Then there exists a constant C such that for every $0 < h \leq h_{\max}$ there exists a stopping index $k(h)$ such that*

$$\sum_{i=1}^{k(h)} H_i \leq C < \infty$$

and $k(h) \to \infty$ as $h \to 0$. Furthermore $\lim_{h \to 0} \frac{1}{\alpha_{k(h)}} \|u_{k(h)}^{in} - u^\dagger\|_{L^2(\Omega)} = 0$.

Proof. We only have to show the existence of such a stopping index. The convergence result then is a direct consequence of Theorem 4.2.1. Let us define the following auxiliary variables

$$A_i := \frac{\rho_i}{\alpha_i} + \frac{\gamma_{i-1}}{\alpha_i} + \frac{\gamma_{i-1}\rho_i}{\alpha_i},$$

$$B_i := \frac{\rho_i^2}{\alpha_i^2} + \frac{\rho_i^2}{\alpha_i}.$$

Now choose $C > 0$ sufficiently large such that

$$\delta(h_{\max})A_1 + \delta(h_{\max})^2 B_1 \leq C.$$

Now pick $0 < h \leq h_{\max}$. Since $\delta : (0, \infty) \to \mathbb{R}$ is a monotonically increasing function function we get the existence of $\tilde{k} \in \mathbb{N}, \tilde{k} \geq 1$ such that

$$\sum_{i=1}^{\tilde{k}} H_i \leq C.$$

Hence, the following expression is well-defined

$$k(h) := \max \left\{ k \in \mathbb{N} : \sum_{i=1}^{k} H_i \leq C \right\}.$$

It is left to show that $k(h) \to \infty$ as $h \to 0$. Assume that this is wrong, hence there exists a $\bar{k} \in \mathbb{N}$ and sequence $(h_n)_n$ with $0 < h_n \leq h_{\max}$ and $h_n \to 0$ such that $k(h_n) < \bar{k}$ holds for all $n \in \mathbb{N}$. By definition we now obtain for all $n \in \mathbb{N}$

$$\delta(h_n) \sum_{i=1}^{\bar{k}} A_i + \delta(h_n)^2 \sum_{i=1}^{\bar{k}} B_i \geq \delta(h_n) \sum_{i=1}^{k(h_n)+1} A_i + \delta(h_n)^2 \sum_{i=1}^{k(h_n)+1} B_i > C.$$

However, since A_i and B_i are independent from h this is a contradiction for n big enough. This finishes the proof. □

This result reflects the expected fact, that a finer discretization leads to a better approximation. In fact we obtain convergence for $h \to 0$, which is not surprising, as for $h \to 0$ Algorithm 3.2 is obtained.

If the discretized subproblem is only solved inexactly we can establish the following result. The proof is a combination of Corollary 4.2.2 and Corollary 4.2.3.

Corollary 4.2.4. *Let $u^\dagger$ satisfy Assumption (SC) and let $(\varepsilon_k)_k$ be a sequence of positive real numbers such that*

$$\sum_{i=1}^{\infty} R_i < \infty.$$

Let $h > 0$ be given. Then there exists a constant C such that for every $0 < h \leq h_{\max}$ there exists a stopping index $k(h)$ such that

$$\sum_{i=1}^{k(h)} H_i \leq C < \infty$$

and $k(h) \to \infty$ as $h \to 0$. Furthermore

$$\lim_{h \to 0} \frac{1}{\alpha_{k(h)}} \| u_{k(h)}^{\text{in}} - u^\dagger \|_{L^2(\Omega)}^2 = 0.$$

4.2.3 Convergence under Active Set Condition

Let us now consider the case when Assumption (ASC) is satisfied.

Theorem 4.2.5. *Let $u^\dagger$ satisfy Assumption (ASC) and let $(\varepsilon_k)_k$ be a sequence of positive real numbers. Furthermore let $h > 0$ be given and let $(u_k^{\text{in}})_k$ be a sequence generated by Algorithm 4.4. Then we have the estimate*

$$\sum_{i=1}^{k} \frac{1}{\alpha_i} \| u_i^{\text{in}} - u^\dagger \|_{L^2(\Omega)}^2 \leq c \left(1 + \sum_{i=1}^{k} \frac{\gamma_{i-1}^{-\kappa}}{\alpha_i} + \sum_{i=1}^{k} R_i + \sum_{i=1}^{k} H_i \right)$$

with the abbreviations

$$R_i := \frac{\varepsilon_i}{\alpha_i} + \frac{\varepsilon_i^2}{\alpha_i^2} + \frac{\gamma_{i-1}\varepsilon_i}{\alpha_i} + \frac{\varepsilon_i^2}{\alpha_i},$$

$$H_i := \delta(h)\left(\frac{\rho_i}{\alpha_i} + \frac{\gamma_{i-1}}{\alpha_i} + \frac{\gamma_{i-1}\rho_i}{\alpha_i}\right) + \delta(h)^2\left(\frac{\rho_i^2}{\alpha_i^2} + \frac{\rho_i^2}{\alpha_i}\right).$$

Proof. The proof mainly follows the idea of Theorem 4.2.1. Again the main part is to establish estimates for the regularization error for u_{k+1}^{ex}. First we want to estimate the term $(u^\dagger, u^\dagger - u)_{L^2(\Omega)}$ using Assumption (ASC). We use Lemma 3.4.19 and obtain

$$(u^\dagger, u^\dagger - u)_{L^2(\Omega)} \le (S^* w, u^\dagger - u)_{L^2(\Omega)} + c\|u^\dagger - u\|_{L^1(A)}, \quad \forall u \in U_{\text{ad}}.$$

This inequality introduces an additional L^1-term. To compensate this term we use the improved optimality condition established in Lemma 3.4.18

$$(-p^\dagger, u - u^\dagger)_{L^2(\Omega)} \ge c_A \|u - u^\dagger\|_{L^1(A)}^{1+\frac{1}{\kappa}}, \quad \forall u \in U_{\text{ad}},$$

with $c_A > 0$. Similar to the proof of Theorem 4.2.1 we start with the following inequality

$$\frac{1}{\alpha_{k+1}}\|S(u_{k+1}^{\text{ex}} - u^\dagger)\|_{L^2(\Omega)}^2 + \|u_{k+1}^{\text{ex}} - u^\dagger\|_{L^2(\Omega)}^2 \le (u^\dagger - \lambda_k^{\text{in}}, u^\dagger - u_{k+1}^{\text{ex}}).$$

Similar to (4.8) we compute

$$(-\lambda_k^{\text{in}}, u^\dagger - u_{k+1}^{\text{ex}})_{L^2(\Omega)}$$

$$\le \sum_{i=1}^k \frac{1}{\alpha_i}(S(u_i^{\text{in}} - u^\dagger), S(u^\dagger - u_{k+1}^{\text{ex}}))_Y$$

$$+ \underbrace{\sum_{i=1}^k \frac{1}{\alpha_i}(Su^\dagger - z, S(u^\dagger - u_{k+1}^{\text{ex}}))_Y}_{\le -c_A\|u^\dagger - u_{k+1}^{\text{ex}}\|_{L^1(A)}^{1+\frac{1}{\kappa}}} + c\gamma_k\delta(h)$$

$$\le \alpha_{k+1}(-v_k^{\text{in}}, v_{k+1}^{\text{in}} - v_k^{\text{in}})_Y - c_A\gamma_k\|u^\dagger - u_{k+1}^{\text{ex}}\|_{L^1(A)}^{1+\frac{1}{\kappa}}$$

$$+ c\gamma_k(\varepsilon_{k+1} + \rho_{k+1}\delta(h)) + c\gamma_k\delta(h).$$

Combining this now yields

$$\frac{1}{\alpha_{k+1}^2}\|S(u_{k+1}^{\text{ex}} - u^\dagger)\|_Y^2 + \frac{1}{\alpha_{k+1}}\|u_{k+1}^{\text{ex}} - u^\dagger\|_{L^2(\Omega)}^2$$

$$\le \frac{1}{\alpha_{k+1}}(u^\dagger, u^\dagger - u_{k+1}^{\text{ex}})_{L^2(\Omega)}$$

$$+ (-v_k^{\text{in}}, v_{k+1}^{\text{in}} - v_k^{\text{in}})_Y - \frac{c_A\gamma_k}{\alpha_{k+1}}\|u^\dagger - u_{k+1}^{\text{ex}}\|_{L^1(A)}^{1+\frac{1}{\kappa}}$$

$$+ c\frac{\gamma_k\varepsilon_{k+1}}{\alpha_{k+1}} + c\delta(h)\left[\frac{\gamma_k\rho_{k+1}}{\alpha_{k+1}} + \frac{\gamma_k}{\alpha_{k+1}}\right].$$

(4.12)

Recall that we have the equality

$$(-v_k^{in}, v_{k+1}^{in} - v_k^{in})_Y = \frac{1}{2}\|v_k^{in}\|_Y^2 - \frac{1}{2}\|v_{k+1}^{in}\|_Y^2 + \frac{1}{2}\|v_{k+1}^{in} - v_k^{in}\|_Y^2.$$

As done in Theorem 4.2.1 we obtain with $\tau > 1$ that

$$\frac{1}{2}\|v_{k+1}^{in} - v_k^{in}\|_Y^2 \leq \frac{1}{2\alpha_{k+1}^2}\left(1 + \frac{1}{\tau}\right)\|S(u_{k+1}^{ex} - u^\dagger)\|_Y^2$$
$$+ \frac{c}{\alpha_{k+1}^2}\left(\varepsilon_{k+1}^2 - \rho_{k+1}^2\delta(h)\right).$$

Using these estimates in (4.12) and performing a summation reveals

$$c_\tau \sum_{i=1}^{k} \frac{1}{\alpha_i^2}\|S(u_{k+1}^{ex} - u^\dagger)\|_Y^2 + \sum_{i=1}^{k} \frac{1}{\alpha_i}\|u_{k+1}^{ex} - u^\dagger\|_{L^2(\Omega)}^2$$

$$+ c_A \sum_{i=1}^{k} \frac{\gamma_{i-1}}{\alpha_i}\|u^\dagger - u_{k+1}^{ex}\|_{L^1(A)}^{1+\frac{1}{\kappa}} + \frac{1}{2}\|v_k^{in}\|_Y^2 \leq \sum_{i=1}^{k} \frac{1}{\alpha_i}(u^\dagger, u^\dagger - u_i^{ex})_{L^2(\Omega)}$$

$$+ c \sum_{i=1}^{k}\left[\frac{\varepsilon_i^2}{\alpha_i^2} + \frac{\gamma_{i-1}\varepsilon_i}{\alpha_i}\right] + c\delta(h) \sum_{i=1}^{k}\left[\frac{\gamma_{i-1}\rho_i}{\alpha_i} + \frac{\gamma_{i-1}}{\alpha_i}\right]$$

$$+ c\delta(h)^2 \sum_{i=1}^{k} \frac{\rho_i^2}{\alpha_i^2}.$$

Now we estimate the term $(u^\dagger, u^\dagger - u_{k+1}^{ex})_{L^2(\Omega)}$ using Young's inequality, see Lemma 3.4.7

$$\sum_{i=1}^{k} \frac{1}{\alpha_i}(u^\dagger, u^\dagger - u_i^{ex})_{L^2(\Omega)} \leq \sum_{i=1}^{k} \frac{1}{\alpha_i}\left[(S^*w, u^\dagger - u_i^{ex})_{L^2(\Omega)} + c\|u^\dagger - u_i^{ex}\|_{L^1(A)}\right]$$

$$= \left(w, \sum_{i=1}^{k} \frac{1}{\alpha_i}S(u^\dagger - u_i^{in})\right)_Y + \left(w, \sum_{i=1}^{k} \frac{1}{\alpha_i}S(u_i^{in} - u_{i,h}^{ex})\right)_Y$$

$$+ \left(w, \sum_{i=1}^{k} \frac{1}{\alpha_i}S(u_{i,h}^{ex} - u_i^{ex})\right)_Y + c\sum_{i=1}^{k} \frac{1}{\alpha_i}\|u^\dagger - u_i^{ex}\|_{L^1(A)}$$

$$\leq \|w\|_Y^2 + \frac{1}{4}\|v_k^{in}\|_Y^2 + c\sum_{i=1}^{k} \frac{\varepsilon_i}{\alpha_i} + c\delta(h)\sum_{i=1}^{k} \frac{\rho_i}{\alpha_i}$$

$$+ \frac{c_A}{2}\sum_{i=1}^{k} \frac{\gamma_{i-1}}{\alpha_i}\|u^\dagger - u_i^{ex}\|_{L^1(A)}^{1+\frac{1}{\kappa}} + c\sum_{i=1}^{k} \frac{\gamma_{i-1}^{-\kappa}}{\alpha_i}.$$

65

Using this estimate we obtain

$$
c_\tau \sum_{i=1}^{k} \frac{1}{\alpha_i^2} \|S(u_{k+1}^{\mathrm{ex}} - u^\dagger)\|_Y^2 + \sum_{i=1}^{k} \frac{1}{\alpha_i} \|u_{k+1}^{\mathrm{ex}} - u^\dagger\|_{L^2(\Omega)}^2
$$

$$
+ \frac{c_A}{2} \sum_{i=1}^{k} \frac{\gamma_{i-1}}{\alpha_i} \|u^\dagger - u_{k+1}^{\mathrm{ex}}\|_{L^1(A)}^{1+\frac{1}{\kappa}} + \frac{1}{4} \|v_k^{\mathrm{in}}\|_Y^2 \le \|w\|_Y^2 + c \sum_{i=1}^{k} \frac{\gamma_{i-1}^{-\kappa}}{\alpha_i}
$$

$$
+ c \sum_{i=1}^{k} \left[\frac{\varepsilon_i^2}{\alpha_i^2} + \frac{\gamma_{i-1}\varepsilon_i}{\alpha_i} + \frac{\varepsilon_i}{\alpha_i} \right] + c\delta(h) \sum_{i=1}^{k} \left[\frac{\rho_i}{\alpha_i} + \frac{\gamma_{i-1}\rho_i}{\alpha_i} + \frac{\gamma_{i-1}}{\alpha_i} \right]
$$

$$
+ c\delta(h)^2 \sum_{i=1}^{k} \frac{\rho_i^2}{\alpha_i^2}.
$$

As in the proof of Theorem 4.2.1 we apply triangular inequality to finish the proof. □

Let us now establish convergence results similar to Corollary 4.2.2 and 4.2.3.

Corollary 4.2.6. *Let $u^\dagger$ satisfy Assumption (ASC) and let $(\varepsilon_k)_k$ be a sequence of positive real numbers such that $\gamma_{i-1}\varepsilon_i \to 0$. Furthermore assume that $S_h = S$ and let $(u_k^{\mathrm{in}})_k$ be a sequence generated by Algorithm 4.4. Then we obtain*

$$
\min_{i=1,..,k} \|u_i^{\mathrm{in}} - u^\dagger\|_{L^2(\Omega)} \to 0
$$

as $k \to \infty$.

Proof. The sequence $(\alpha_k)_k$ is bounded by a constant M. Hence we have the following inequalities for k large enough

$$
\sum_{i=2}^{k} \frac{\gamma_{i-1}}{\alpha_i} \varepsilon_i \ge \frac{1}{M} \sum_{i=2}^{k} \frac{i-1}{\alpha_i} \varepsilon_i \ge \frac{1}{M} \sum_{i=2}^{k} \frac{\varepsilon_i}{\alpha_i}.
$$

Furthermore we have with Lemma 3.4.6 that

$$
\gamma_k^{-1} \sum_{i=1}^{k} \frac{\gamma_{i-1}^{-\kappa}}{\alpha_i} + \gamma_k^{-1} \sum_{i=1}^{k} \frac{\gamma_{i-1}}{\alpha_i} \varepsilon_i \to 0.
$$

We now obtain

$$
\min_{i=1,..,k} \|u_i^{\mathrm{in}} - u^\dagger\|_{L^2(\Omega)} \le c \left(\gamma_k^{-1} + \gamma_k^{-1} \sum_{i=1}^{k} \frac{\gamma_{i-1}^{-\kappa}}{\alpha_i} + \gamma_k^{-1} \sum_{i=1}^{k} R_i \right) \to 0,
$$

which finishes the proof. □

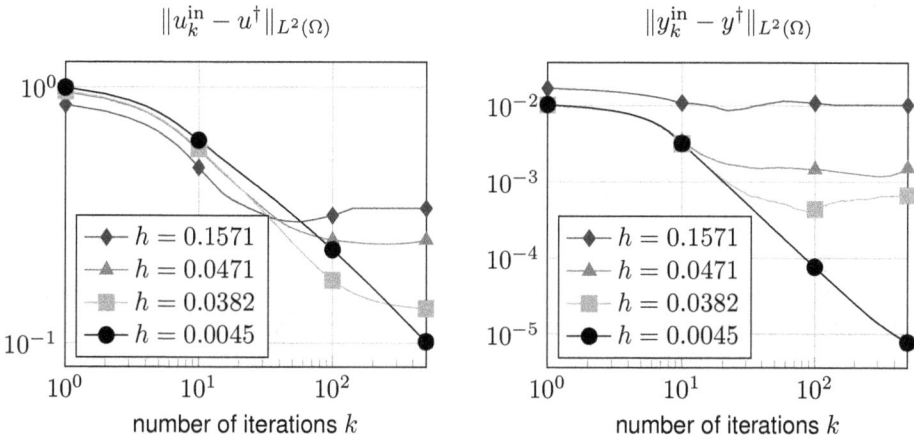

$\|u_k^{\text{in}} - u^\dagger\|_{L^2(\Omega)}$ (left plot y-axis label)

$\|y_k^{\text{in}} - y^\dagger\|_{L^2(\Omega)}$ (right plot y-axis label)

Left plot legend:
- $h = 0.1571$
- $h = 0.0471$
- $h = 0.0382$
- $h = 0.0045$

number of iterations k

Right plot legend:
- $h = 0.1571$
- $h = 0.0471$
- $h = 0.0382$
- $h = 0.0045$

number of iterations k

Figure 4.1: Error $\|u_k^{\text{in}} - u^\dagger\|_{L^2(\Omega)}$ (left) and $\|y_k^{\text{in}} - y^\dagger\|_{L^2(\Omega)}$ (right) for different mesh sizes h.

Comparing this result to Corollary 4.2.2 we see, that we do not obtain strong convergence of the sequence $(u_k^{\text{in}})_k$. But we obtain convergence of the iterates generated by Algorithm 3.2 under the Assumption (ASC). The reason is, that the proof uses a different technique. The proof of Theorem 3.4.20 uses Bregman distances and the fact that $\lambda_k \in \partial J(u_k)$. However, this is not the case here, so the proof cannot be directly copied. Numerical observations, however, still show convergence of the sequence $(u_k^{\text{in}})_k$.

Corollary 4.2.7. *Let $u^\dagger$ satisfy Assumption (ASC). Let $h_{\max} > 0$ be given and $\varepsilon_k = 0$ for all $k \in \mathbb{N}$. Then there exists a constant C such that for every $0 < h \le h_{\max}$ there exists a stopping index $k(h)$ such that*

$$\sum_{i=1}^{k(h)} H_i \le C < \infty$$

and $k(h) \to \infty$ as $h \to 0$. Furthermore

$$\min_{i=1,..,k(h)} \|u_i^{\text{in}} - u^\dagger\|_{L^2(\Omega)} \to 0,$$

as $h \to 0$.

Proof. The proof is very similar to the proof of Corollary 4.2.3. □

A combination of both results yields the following corollary.

Corollary 4.2.8. *Let $u^\dagger$ satisfy Assumption (ASC) and let $(\varepsilon_k)_k$ be a sequence of positive real numbers such that $\gamma_{i-1}\varepsilon_i \to 0$. Let $h > 0$ be given. Then there exists a constant C such that for every $0 < h \le h_{\max}$ there exists a stopping index $k(h)$ such that*

$$\sum_{i=1}^{k(h)} H_i \le C < \infty$$

and $k(h) \to \infty$ as $h \to 0$. Furthermore

$$\min_{i=1,..,k(h)} \|u_i^{in} - u^\dagger\|_{L^2(\Omega)} \to 0$$

as $h \to 0$.

To test the inexact iterative Bregman method (Algorithm 4.4) we consider the example presented in Subsection 5.2.2. We use the implementation presented in Chapter 5.

We use different mesh sizes for comparison and plot the error for the first 500 iterations in Figure 4.1. Here, we set $\alpha_k := 0.1$ and $\varepsilon_k := k^{-3/2}$ to satisfy the assumptions of Corollary 4.2.8. As expected we see that for $h \to 0$ we obtain convergence for $k \to \infty$. The coarsest mesh has approximately 10^2 and the finest mesh approximately 10^5 degrees of freedom.

CHAPTER 5

Implementation of the Iterative Bregman Method

In Chapter 3 we considered an objective functional of the form $\|Su - z\|_Y^2$. Here Y is a Hilbert space, $\Omega \subset \mathbb{R}^n$ a bounded domain and $S : L^2 \to Y$ a linear operator. This operator was replaced in Chapter 4 by a suitable approximation S_h. In this chapter we want to specify this operator. To be precise we consider the linear elliptic partial differential equation

$$-\Delta y = u \text{ in } \Omega,$$
$$y = 0 \text{ on } \partial\Omega,$$

and its numerical approximation using linear finite elements, see also Subsection 2.2.2.

We start by formulating a semi-smooth Newton method in Section 5.1 to solve the subproblems. In Subsection 5.1.1 we analyse the discretization using finite elements. There we also define the operator S_h used in Chapter 4. The chapter is closed in Section 5.2, where we present several numerical example, to show the efficiency of our algorithm and to support the theoretical findings. The results of this chapter can be found in the publications [75, 76].

5.1 Semi-Smooth Newton Method

In our algorithm we need to solve the problem (3.8) in each step, which includes the Bregman distance. Similar to (4.1) and (3.6) this can be written in the following form. For brevity we set $\alpha := \alpha_k$ and $\lambda := \lambda_{k-1} \in L^2(\Omega)$.

$$\text{Minimize} \quad \frac{1}{2}\|Su - z\|_Y^2 + \alpha \left[\frac{1}{2}\|u\|_{L^2(\Omega)}^2 - (\lambda, u)_{L^2(\Omega)} \right] \tag{5.1}$$

$$\text{such that} \quad u \in U_{\text{ad}}.$$

Note that (5.1) has has a unique solution, characterized by the projection formula

$$u = P_{U_{\text{ad}}} \left(-\frac{1}{\alpha} p(u) + \lambda \right), \tag{5.2}$$

with $p(u) = S^*(Su - z)$. Several different techniques are available to solve (5.1).
The simplest is a projected gradient method, see [93], with the descent direction
$-p(u) - \alpha(u - \lambda)$. The implementation is rather simple but comes at very slow
convergence speed and high numerical costs. Nevertheless the gradient method can
be used to globalize the Newton method presented below.

In order to solve (5.1) we want to apply a semi-smooth Newton method to (5.2).
In this section we follow the idea presented in [47], where a semi-smooth Newton
solver was applied for a Neumann-type elliptic optimal control problem. We adapt
this technique for distributed control problems. This technique can also be applied
for optimal boundary control problems, see [6]. Denote by u^k the iterates given by
the Newton method. Define the function

$$F(u) := u - P_{U_{\mathrm{ad}}}\left(-\frac{1}{\alpha}p(u) + \lambda\right).$$

The function F is semi-smooth and we can apply a Newton step

$$0 = F(u^k) + G(u^k)(u^{k+1} - u^k),$$

where $G(u^k) : L^2(\Omega) \to L^2(\Omega)$ is a suitable derivative of F at u^k. To be precise G has
to satisfy a superlinear approximation condition, for more details see [48, Section
2.4.4]. For a convergence analysis of this Newton method we refer to [47]. A suitable
function G is given by the following lemma. The result can also be found in [6] or
in [48, Theorem 2.14].

Lemma 5.1.1. *A suitable derivative $G(u^k) : L^2(\Omega) \to L^2(\Omega)$ of F at u_k is given by*

$$G(u^k)(u^{k+1} - u^k) = (u^{k+1} - u^k) + d \cdot \left(\frac{1}{\alpha}\left(p(u^{k+1}) - p(u^k)\right)\right)$$

with

$$d = \begin{cases} 0, & \text{if } -\frac{1}{\alpha}p(u^k) + \lambda \geq u_b, \\ 1, & \text{if } -\frac{1}{\alpha}p(u^k) + \lambda \in (u_a, u_b), \\ 0, & \text{if } -\frac{1}{\alpha}p(u^k) + \lambda \leq u_a. \end{cases}$$

We see that u^{k+1} satisfies the relation

$$u^{k+1} = \begin{cases} u_b, & \text{if } -\frac{1}{\alpha}p(u^k) + \lambda \geq u_b, \\ u_a, & \text{if } -\frac{1}{\alpha}p(u^k) + \lambda \leq u_a, \\ -\frac{1}{\alpha}p(u^{k+1}) + \lambda, & \text{if } -\frac{1}{\alpha}p(u^k) + \lambda \in (u_a, u_b). \end{cases}$$

Define the sets

$$A_b(u) = \left\{x \in \Omega : -\frac{1}{\alpha}p(u) + \lambda \geq u_b\right\},$$

$$I(u) = \left\{x \in \Omega : -\frac{1}{\alpha}p(u) + \lambda \in (u_a, u_b)\right\},$$

$$A_a(u) = \left\{x \in \Omega : -\frac{1}{\alpha}p(u) + \lambda \leq u_a\right\},$$

and the operator

$$E_M : L^2(\Omega) \to L^2(\Omega), \quad E_M(v) = \chi_M(v).$$

All the inequalities are to be understood pointwise almost everywhere. We have $u^{k+1} = u_a$ on $A_a(u^k)$ and $u^{k+1} = u_b$ on $A_b(u^k)$. On the set $I(u^k)$ we obtain

$$E_{I(u^k)}\left(u^{k+1} + \frac{1}{\alpha}p(u^{k+1}) - \lambda\right) = 0.$$

This can be rewritten in a linear equation for $E_{I(u^k)}u^{k+1}$.

Lemma 5.1.2. *The function* $u_I^{k+1} := E_{I(u^k)}u^{k+1}$ *satisfies*

$$u_I^{k+1} + \frac{1}{\alpha}E_{I(u^k)}q(u_I^{k+1}) = -\frac{1}{\alpha}E_{I(u^k)}p\big(E_{A_a(u^k)}u_a + E_{A_b(u^k)}u_b\big) + E_{I(u^k)}\lambda, \quad (5.3)$$

with $q(u) := S^*Su$.

Our Newton solver now solves the equation above for u_I^{k+1}, which allows us to construct our new iterate u^{k+1}. Note, that we have formulated the Newton method in an abstract Banach space setting. When we apply a finite element discretization we will test (5.3) with some test functions to obtain a finite-dimensional system. This is part of the next subsection.

5.1.1 Algorithmic Aspects and Implementation

We now focus on the special case where $y = Su$ is given by the weak solution of the linear elliptic partial differential equation for a convex domain $\Omega \subset \mathbb{R}^n$ with $n = 2, 3$.

$$\begin{aligned}-\Delta y &= u \quad \text{in } \Omega, \\ y &= 0 \quad \text{on } \partial\Omega.\end{aligned} \qquad (5.4)$$

Let us show that this example fits into our framework. Clearly, for $u \in L^2(\Omega)$ equation (5.4) has a unique weak solution $y \in H_0^1(\Omega)$, and the associated solution operator S is linear and continuous. For the choice $Y = L^2(\Omega)$ we obtain $S^* = S$.

Let us now report on the discretization and the operator S_h. We follow the argumentation and results presented in [96, Section 3]. Let $\mathcal{T}_h$ be a regular mesh which consists of closed cells T. We assume that the union of all cells is the whole domain Ω. For $T \in \mathcal{T}_h$ we define $h_T := \operatorname{diam} T$. Furthermore we set $h := \max_{T \in \mathcal{T}_h} h_T$. We assume that there exists a constant $R > 0$ such that $\frac{h_T}{R_T} \leq R$ for all $T \in \mathcal{T}$. Here we define R_T to be the diameter of the largest ball contained in T.

For this mesh $\mathcal{T}$ we define an associated finite dimensional space $Y_h \subset H_0^1(\Omega)$, such that the restriction of a function $v \in Y_h$ to a cell $T \in \mathcal{T}$ is a linear polynomial. The operator S_h is now defined in the sense of weak solutions. We set $y_h := S_h u$ if $y_h \in Y_h$ solves

$$a(y_h, v_h) = (u, v_h)_{L^2(\Omega)} \quad \forall v_h \in Y_h,$$

with the bilinear form

$$a(w, v) := \int_{\Omega} \nabla w \cdot \nabla v \; dx.$$

We also obtain $S_h^* = S_h$ in the discrete case. Let us now mention that the operator S_h satisfies Assumption (4.1.1). Following [96] and the references therein we obtain the following result.

Lemma 5.1.3. *Assume that there exists a constant $C_M > 1$ such that*

$$\max_{T \in \mathcal{T}_h} h_T \leq C_M \min_{T \in \mathcal{T}_h} h_T$$

holds. Then we have the estimates

$$\|(S - S_h)f\|_{L^2(\Omega)} \leq ch^2 \|f\|_{L^2(\Omega)},$$
$$\|(S^* - S_h^*)f\|_{L^\infty(\Omega)} \leq ch^{2-n/2} \|f\|_{L^2(\Omega)},$$

for $f \in L^2(\Omega)$ and a constant c independent from f and h.

Hence Assumption (4.1.1) is satisfied with $\delta(h) = ch^2$. The discretized version of (5.1) is now given by the solution (u_h, y_h, p_h) of the coupled problem

$$
\begin{aligned}
a(y_h, v_h) &= (u_h, v_h), &\forall v_h \in Y_h, \\
a(p_h, v_h) &= (v_h, y_h - z_h), &\forall v_h \in Y_h, \\
u_h &= P_{U_{\mathrm{ad}}}\left(-\frac{1}{\alpha}p_h + \lambda_h\right),
\end{aligned}
\tag{5.5}
$$

with $\lambda_h \in Y_h$. For a given u_h there exists a unique $y_h(u_h)$ and hence a unique $p_h(u_h)$, so we reduce the coupled system (5.5) to one equation for the optimal control u_h

$$u_h = P_{U_{\mathrm{ad}}}\left(-\frac{1}{\alpha}p_h(u_h) + \lambda_h\right).$$

Note that Lemma 5.1.2 also holds for finite element functions. We are interested in the solution $u_{h,I}^{k+1}$ from equation (5.2). But $u_{h,I}^{k+1}$ is not a finite element function in general, since it is the truncation of a finite element function $u_{h,I}^{k+1} = E_{I(u_h^k)}\tilde{u}_h^{k+1}$, which can be computed by solving the following equation, see (5.3)

$$
\begin{aligned}
E_{I(u_h^k)}\tilde{u}_h^{k+1} &+ \frac{1}{\alpha}E_{I(u_h^k)}q\left(E_{I(u_h^k)}\tilde{u}_h^{k+1}\right) \\
&= -\frac{1}{\alpha}E_{I(u_h^k)}p(E_{A_a(u_h^k)}u_a + E_{A_b(u_h^k)}u_b) + E_{I(u_h^k)}\lambda_h.
\end{aligned}
\tag{5.6}
$$

In the following we denote by $\underline{u_h} \in \mathbb{R}^m$ the coefficient vector of a function $u_h \in Y_h$, where m denotes the degrees of freedom. By testing (5.6) with a test function we obtain the following lemma.

Lemma 5.1.4. *The coefficient vector* $\underline{\tilde{u}_h^{k+1}}$ *satisfies*

$$\left(M_I + \frac{1}{\alpha} M_I K^{-1} M K^{-1} M_I\right)\underline{\tilde{u}_h^{k+1}} = -\frac{1}{\alpha} M_I K^{-1} M\left(K^{-1}g - \underline{z}\right) + M_I \underline{\lambda_h}, \quad (5.7)$$

where

$$K = \left[\int_\Omega \nabla\varphi_i \cdot \nabla\varphi_j\right]_{ij},$$

$$M_I = \left[\int_{I(u_h^k)} \varphi_i\varphi_j\right]_{ij}, \quad M_{A_a} = \left[\int_{A_a(u_h^k)} \varphi_i\varphi_j\right]_{ij},$$

$$M = \left[\int_\Omega \varphi_i\varphi_j\right]_{ij}, \quad M_{A_b} = \left[\int_{A_b(u_h^k)} \varphi_i\varphi_j\right]_{ij},$$

$$g = \left[\int_{A_a(u_h^k)} u_a\varphi_j + \int_{A_b(u_h^k)} u_b\varphi_j\right]_j = M_{A_a}\underline{u_a} + M_{A_b}\underline{u_b}.$$

Note that we now have the relation

$$u_h^{k+1} = E_{A_a(u_h^k)}u_a + E_{A_b(u_h^k)}u_b + E_{I(u_h^k)}\tilde{u}_h^{k+1}.$$

We can use this relation to get a system for the coefficient vector of the function p_h^{k+1}.

Lemma 5.1.5. *The coefficient vector of the adjoint state* p_h^{k+1} *satisfies*

$$\underline{p_h^{k+1}} = K^{-1}M\left(K^{-1}(g + M_I\underline{\tilde{u}_h^{k+1}}) - \underline{z_h}\right).$$

Note that only the adjoint state p_h^{k+1} is used to update the active and inactive sets, hence kinks and discontinuities will not be accumulated.

As mentioned in [6] the operator on the left-hand side of (5.3) is positive definite on $L^2(I(u^k))$, hence the matrix on the left-hand side of (5.7) is positive definite on the span of all basis functions whose support has non-empty intersection with the inactive set $I(u_h^k)$. This makes the equation accessible with a conjugate gradient method.

With these results we can implement our Newton method and solve the subproblem without actually computing u_h^k, we only work with adjoint state and the active and inactive sets.

5.1.2 Using the Newton Solver in the Bregman Iteration

The fact that we are not computing the control (which is not an FEM function) and work instead with the adjoint state (which is an FEM function) can be extended to the implementation of the iterative Bregman method. Denote k the number of iterations and let $\lambda_h^k \in Y_h$ be the computed subgradient. Let $p_h^{k+1} \in Y_h$ be the adjoint state computed while solving the subproblem. To start the next iteration all we have to do is to update the subgradient

$$\lambda_h^{k+1} := -\frac{1}{\alpha_{k+1}} p_h^{k+1} + \lambda_h^k \in Y_h.$$

Again note that we do not need to compute the control. The control can be computed (for plotting e.g.) using the optimality condition $u_h^{k+1} = P_{U_{\mathrm{ad}}}(\lambda_h^{k+1})$ if needed.

5.2 Numerical Results

In this section we present several different examples to illustrate the efficiency of Algorithm 3.2. We now consider the following optimal control problem. Note that due to the linearity of S this is of form (P).

$$\text{Minimize} \quad \frac{1}{2}\|y - z\|_{L^2(\Omega)}^2$$

$$\text{such that} \quad -\Delta y = u + e_\Omega \quad \text{in } \Omega,$$

$$y = 0 \quad \text{on } \partial\Omega, \tag{5.8}$$

$$u_a \leq u \leq u_b \quad \text{a.e. in } \Omega.$$

The additional variable e_Ω can be used to construct the optimal solution of (5.8). For the construction process we refer to [93]. The implementation of the semi-smooth Newton method was done with FEniCS [64] using the C++ interface.

5.2.1 One-Dimensional Examples

Let us start with $\Omega \subset \mathbb{R}$. In this case it is possible to construct optimal solutions $(u^\dagger, y^\dagger, p^\dagger)$, such that $u^\dagger$ exhibit a bang-bang structure in one part of the domain Ω, and is singular, i.e. $p^\dagger = 0$ on the other part. We use the implementation presented in Section 5.1 with an equidistant subdivision of the interval Ω with mesh size h.

Example 1: Source Condition

We want to construct an optimal control such that $u^\dagger = S^* w$ for some $w \in L^2(\Omega)$. This is a stronger version of the assumption made in Assumption (SC). Note that in this case also Assumption (SC) is satisfied.

Let $\Omega = (-1, 1)$, $u_a = -1$ and $u_b = 1$. We now define

$$u^\dagger(x) = \frac{\pi^2}{10}\sin(\pi x),$$

$$y^\dagger(x) = \frac{1}{10}\sin(\pi x),$$

$$p^\dagger(x) = 0,$$

$$e_\Omega(x) = 0,$$

$$z(x) = y^\dagger(x).$$

It is easy to check that the functions $(u^\dagger, y^\dagger, p^\dagger)$ are a solution to (5.8). The desired state z is reachable by construction. Furthermore we have $-\Delta u = -\pi^2 u^\dagger =: w$. This shows that the Source Condition (SC) is satisfied. We set $\alpha_k = 1$ and compute the first 1000 iterations with mesh size $h = 4 \cdot 10^{-6}$. The computed errors can be seen in Figure 5.4.

Example 2: Projected Source Condition

In this example we want to construct an optimal control problem such that for the optimal control $u^\dagger = P_{U_{ad}}(S^*w)$ holds with some $w \in L^2(\Omega)$. This is related to Example 1, where $u^\dagger = S^*w$ holds. Again let $\Omega = (-1,1)$, $u_a = 0$ and $u_b = 1$. With the choice of

$$u^\dagger(x) = \begin{cases} (x+1)(x+\tfrac{1}{2})(x-\tfrac{1}{2})(x-1) & \text{if } x \in [-\tfrac{1}{2},\tfrac{1}{2}], \\ 0 & \text{else,} \end{cases}$$

$$y^\dagger(x) = \begin{cases} \frac{19}{240}(x+1) & \text{if } x \in [-1,-\tfrac{1}{2}], \\ \frac{19}{240}(x+1) - \frac{11}{768} - \frac{19x}{240} - \frac{x^2}{8} + \frac{5x^4}{48} - \frac{x^6}{30} & \text{if } x \in [-\tfrac{1}{2},\tfrac{1}{2}], \\ \frac{19}{240}(x+1) - \frac{19}{240} - \frac{19}{120}(x-\tfrac{1}{2}) & \text{if } x \in [\tfrac{1}{2},1], \end{cases}$$

$$p^\dagger(x) = 0,$$
$$z(x) = y^\dagger(x) - \Delta p^\dagger(x),$$
$$e_\Omega(x) = 0$$

the functions $(u^\dagger, y^\dagger, p^\dagger)$ are a solution to (5.8). The functions are plotted in Figure 5.2. Furthermore with the choice of

$$w(x) = \frac{5}{2} - 12x^2 = -\Delta\left((x+1)(x+\tfrac{1}{2})(x-\tfrac{1}{2})(x-1)\right)$$

we obtain $u^\dagger = P_{U_{ad}}(S^*w)$. Again we fix $\alpha_k = 0.1$ and compute the first 2048 iterations with mesh size $h = 4 \cdot 10^{-7}$ leading to 10^7 degrees of freedom. The computed errors can be seen in Figure 5.5.

Example 3: Active Set Condition

We now want to construct an example, where the optimal control $u^\dagger$ has a bang-bang structure in one part of the domain, and is singular on the other. The regularity assumption (ASC) is designed for such a structure. The construction of the bang-bang part is straightforward and for the part where the solution is singular we define $u^\dagger$ to be continuous. The optimal state is now obtained by solving the resulting partial differential equation. The optimal adjoint state is now chosen such that the pair $(u^\dagger, p^\dagger)$ is consistent in the sense of Theorem 3.2.2. Finally we construct the desired state such that $(u^\dagger, y^\dagger, p^\dagger)$ are a solution of (5.8).

We set $\Omega = (-1, 1)$, $u_a = 0$, $u_b = \frac{1}{10}$ and

$$u^\dagger(x) = \begin{cases} \frac{1}{10} & \text{if } x \in [-1, -\frac{1}{2}], \\ 0 & \text{if } x \in [-\frac{1}{2}, \frac{1}{4}], \\ (x+1)(x-\frac{1}{4})(x-\frac{3}{4})(x-1) & \text{if } x \in [\frac{1}{4}, \frac{3}{4}], \\ 0 & \text{if } x \in [\frac{3}{4}, 1], \end{cases}$$

$$y^\dagger(x) = \begin{cases} -\frac{7}{3072} - \frac{803x}{15360} - \frac{x^2}{20} & \text{if } x \in [-1, -\frac{1}{2}], \\ \frac{157}{15360} - \frac{7x}{3072} & \text{if } x \in [-\frac{1}{2}, \frac{1}{4}], \\ \frac{581}{49152} - \frac{11x}{480} + \frac{3x^2}{32} - \frac{x^3}{6} + \frac{13x^4}{192} + \frac{x^5}{20} - \frac{x^6}{30} & \text{if } x \in [\frac{1}{4}, \frac{3}{4}], \\ \frac{271}{15360} - \frac{271x}{15360} & \text{if } x \in [\frac{3}{4}, 1], \end{cases}$$

$$p^\dagger(x) = \begin{cases} -(x+1)(x+\frac{1}{2})^4 & \text{if } x \in [-1, -\frac{1}{2}], \\ -30(x+\frac{1}{2})^4 x^3 & \text{if } x \in [-\frac{1}{2}, 0], \\ 0 & \text{if } x \in [0, 1], \end{cases}$$

$$z(x) = y^\dagger(x) - \Delta p^\dagger(x),$$
$$e_\Omega(x) = 0.$$

Hence, the functions $(u^\dagger, y^\dagger, p^\dagger)$ are a solution to (5.8). A plot of these functions can be found in Figure 5.3. On the interval $I = (0, 1) \subset \Omega$ we have that $u^\dagger = P_{U_{\mathrm{ad}}}(S^*w)$ holds, see Example 2. On the interval $A := \Omega \setminus I$, the control $u^\dagger$ has a bang-bang structure, i.e. $u^\dagger(x) \in \{u_a(x), u_b(x)\}$ almost everywhere. Using Theorem 3.3.6 yield that Assumption (ASC) is satisfied with $\kappa = \frac{1}{4}$. We compute the first 2048 iterations for fixed $\alpha_k = 0.1$ with mesh size $h = 10^{-6}$. From theory we expect $\|u_k - u^\dagger\|_{L^2(\Omega)} = \mathcal{O}\left(k^{-\frac{1}{8}}\right)$. The computed errors can be seen in Figure 5.6.

Example 4: Bang-Bang Solution with $\kappa = 1$

Next we construct an example where $u^\dagger$ exhibits a bang-bang structure on Ω. We set $\Omega = (-1, 1)$, $u_a = -1$, $u_b = 1$ and

$$p^\dagger(x) = \sin(\pi x),$$
$$u^\dagger(x) = -\mathrm{sign}(p^\dagger),$$
$$y^\dagger(x) = 1 - x^2,$$
$$e_\Omega(x) = -\Delta y^\dagger(x) - u^\dagger(x),$$
$$z(x) = y^\dagger(x) + \Delta p^\dagger(x).$$

The functions $(u^\dagger, y^\dagger, p^\dagger)$ are a solution to (5.8). By Theorem 3.3.4 or 3.3.6 Assumption (ASC) is satisfied with $A = \Omega$ and $\kappa = 1$.

It is known, that Assumption (ASC) is not only sufficient, but also necessary for some regularization error estimates obtained for the Tikhonov regularization for problem (P), see [98]. We expect that a similar result also holds for our algorithm. However, it is an open problem to prove this relation.

We fix $\alpha_k = 1$ and compute the first 2048 iterations with mesh size $h = 4 \cdot 10^{-7}$. We expect a convergence rate $\|u_k - u^\dagger\|_{L^2(\Omega)}^2 = \mathcal{O}\left(k^{-1}\log(k)\right)$ see Corollary 3.4.5. We also compute the rate

$$\kappa_k := \frac{1}{\log(2)}\log\left(\frac{\|u_{k/2} - u^\dagger\|_{L^2(\Omega)}^2}{\|u_k - u^\dagger\|_{L^2(\Omega)}^2}\right).$$

The results for the numerical convergence rate κ_k can be found in Table 5.1 and the computed errors in Figure 5.7. We see that $\kappa_k \approx 1$, which supports the hypothesis, that Assumption (ASC) is also necessary. Note that for larger k the discretization error is dominating, leading to unreliable results for κ_k.

	Example 4		Example 5	
k	$\|u_k - u^\dagger\|_{L^2(\Omega)}$	κ_k	$\|u_k - u^\dagger\|_{L^2(\Omega)}$	κ_k
1	0.6722687597530747	-	0.7259004437566887	-
2	0.4632648773518065	1.0744	0.5749379045091544	0.6727
4	0.3260939576765771	1.0131	0.4202572744669416	0.9043
8	0.2303595953689695	1.0028	0.3210829677835648	0.7767
16	0.1628606909423222	1.0005	0.2519388899120427	0.6997
32	0.1151597670894625	1.0000	0.200898312209989	0.6532
64	0.0814343040205087	0.9999	0.1622230718552173	0.6170
128	0.05759141154404519	0.9996	0.1323174230010808	0.5880
256	0.04074577258246533	0.9984	0.1085464267526645	0.5714
512	0.02887391254287635	0.9938	0.08997521562370121	0.5414
1024	0.02059120915374126	0.9755	0.07475963451887319	0.5345
2048	0.01504185572910729	0.9061	0.06232732655147037	0.5248

Table 5.1: Computed rate κ_k for Example 4 and Example 5.

Example 5: Bang-Bang Solution with $\kappa = \frac{1}{2}$

We define the following functions on $\Omega = (0,1)$:

$$p^\dagger = x^2(x-1)(2x-1),$$
$$u^\dagger = -\text{sign}(p^\dagger),$$
$$y^\dagger = x(1-x),$$
$$e_\Omega(x) = -\Delta y^\dagger(x) - u^\dagger(x),$$
$$z(x) = y^\dagger(x) + \Delta p^\dagger(x).$$

The functions $(u^\dagger, y^\dagger, p^\dagger)$ are a solution to (5.8). Clearly $u^\dagger$ has a bang-bang structure. Here Theorem 3.3.6 reveals that Assumption (ASC) is satisfied with $A = \Omega$ and $\kappa = \frac{1}{2}$. Similar to Example 4, we fix $\alpha_k = 0.1$ and $h = 2 \cdot 10^{-7}$. The computed errors can be seen in Figure 5.8. We expect a convergence rate $\|u_k - u^\dagger\|_{L^2(\Omega)}^2 = \mathcal{O}(k^{-\frac{1}{2}})$, which is good approximated, see Table 5.1.

Example 6: Bang-Bang Solution with $\kappa = \frac{1}{3}$

Similar to Example 5 we define on $\Omega = (0, 1)$ the functions

$$p^\dagger = x(1 - x)(3x - 1)^3,$$
$$u^\dagger = -\text{sign}(p^\dagger),$$
$$y^\dagger = x(1 - x),$$
$$e_\Omega(x) = -\Delta y^\dagger(x) - u^\dagger(x),$$
$$z(x) = y^\dagger(x) + \Delta p^\dagger(x).$$

The functions $(u^\dagger, y^\dagger, p^\dagger)$ are a solution to (5.8). A calculation using Theorem 3.3.6 shows that Assumption (ASC) is satisfied with $A = \Omega$ and $\kappa = \frac{1}{3}$. The computed results for κ_k can be found in Table 5.2. The expected value $\kappa = \frac{1}{3}$ is obtained. Here we use $h = 2 \cdot 10^{-7}$.

	Example 6		Example 8	
k	$\|u_k - u^\dagger\|_{L^2(\Omega)}$	κ_k	$\|u_k - u^\dagger\|_{L^2(\Omega)}$	κ_k
1	0.6384310006585554	-	0.9478136297478046	-
2	0.5564941402566002	0.3963	0.897976145981608	0.1559
4	0.4736649311926216	0.4650	0.8047502704210617	0.3163
8	0.4138390253183378	0.3896	0.655737352082853	0.5908
16	0.3646955085938504	0.3648	0.5093855712942857	0.7287
32	0.322728036963586	0.3527	0.3901466377095458	0.7695
64	0.2862541886157004	0.3460	0.2951825660864547	0.8048
128	0.2542396065205812	0.3422	0.2208871361735489	0.8366
256	0.2259682413252065	0.3401	0.1634645681003727	0.8687
512	0.2009017117358534	0.3393	0.1193857565988463	0.9067
1024	0.1786060483761079	0.3394	0.08549473690142423	0.9634
2048	0.1587149243160397	0.3407	0.0601051288801563	1.0167

Table 5.2: Numerical convergence rate κ_k for Example 6 and Example 8.

5.2.2 Two-Dimensional Examples

In this Subsection we want to present test examples for $\Omega \subset \mathbb{R}^2$. In contrast to the one-dimensional case, it is not clear how to construct examples similar to Example 3. However, we will present some interesting examples. We use a regular triangulation of the domain Ω with mesh size h, as presented in Subsection 5.1.1.

Example 7: Source Condition in 2D

For $\Omega = (0,1)^2$, $u_a = -1$, $u_b = 1$ we define

$$u^\dagger(x,y) = \sin(\pi x)\sin(\pi y),$$
$$y^\dagger(x,y) = \frac{1}{2\pi^2}\sin(\pi x)\sin(\pi y),$$
$$p^\dagger(x,y) = 0,$$
$$e_\Omega(x), y = 0,$$
$$z(x,y) = y^\dagger(x,y).$$

Similar to Example 1 the desired state is reachable and the source condition (SC) is satisfied. We use a regular triangular mesh with $h \approx 1.4 \cdot 10^{-3}$, leading to approximately $3.3 \cdot 10^5$ degrees of freedom and compute the first 1000 iterates. The numerical results can be seen in Figure 5.10.

Example 8: Bang-Bang Solution with $\kappa = 1$ in 2D

With the choice of $\Omega = (0,1)^2$, $u_a = -1$, $u_b = 1$ and

$$p^\dagger(x,y) = -\frac{1}{8\pi^2}\sin(2\pi x)\sin(2\pi y),$$
$$u^\dagger(x,y) = -\text{sign}(p^\dagger(x)),$$
$$y^\dagger(x,y) = \sin(\pi x)\sin(\pi y),$$
$$e_\Omega(x,y) = 2\pi^2\sin(\pi x)\sin(\pi y) - u^\dagger,$$
$$z(x,y) = \sin(\pi x)\sin(\pi y) + \sin(2\pi x)\sin(2\pi y)$$

the functions $(u^\dagger, y^\dagger, p^\dagger)$ are a solution to (5.8). The optimal control is of bang-bang structure, i.e. $u^\dagger(x) \in \{-1, 1\}$ almost everywhere, however it is not clear whether Assumption (ASC) is satisfied. The assumptions of Theorem 3.3.4 are not satisfied (consider the point $(\frac{1}{2}, \frac{1}{2})$), but numerical investigations indicate that Assumption (ASC) is satisfied with $A = \Omega$ and $\kappa = 1$. This is supported by the computed regularization error rates and by a direct computation of κ for the function $p^\dagger$ on a small grid.

We use a regular triangular mesh with $h \approx 1.4 \cdot 10^{-3}$, leading to approximately $3.3 \cdot 10^5$ degrees of freedom. We fix $\alpha_k = 0.1$. The obtained errors can be seen in Figure 5.11 and the numerical convergence order is computed in Table 5.2. We see that $\kappa_k \approx 1$ which is expected by the theoretical findings.

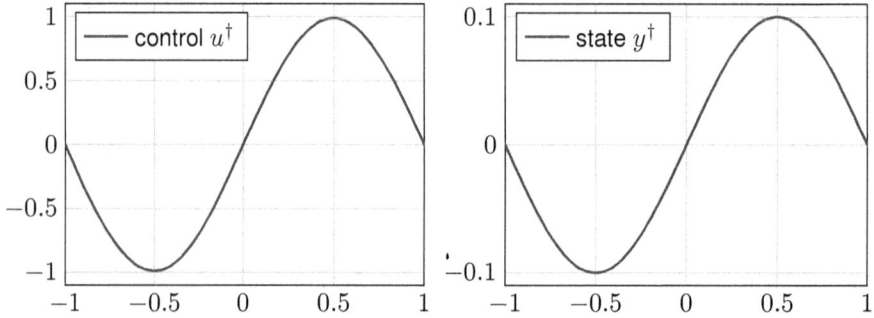

Figure 5.1: Optimal control $u^\dagger$ (left) and state $y^\dagger$ (right) for Example 1.

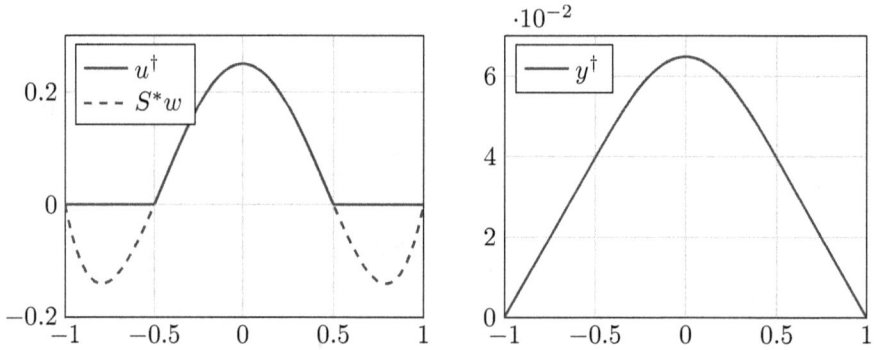

Figure 5.2: Optimal control $u^\dagger = P_{U_{ad}}(S^*w)$ (left) and the optimal state $y^\dagger$ (right) for Example 2.

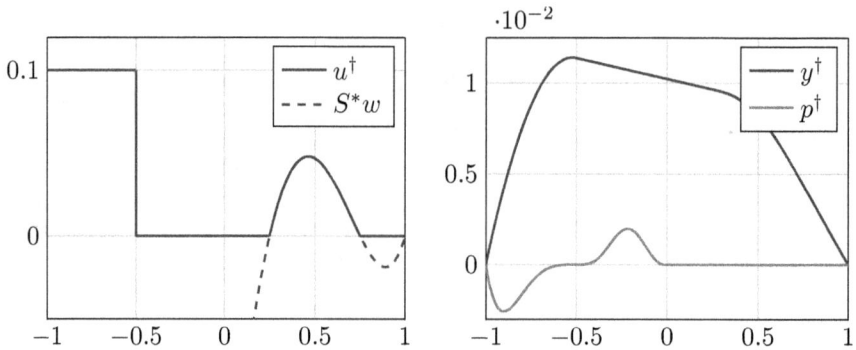

Figure 5.3: Optimal control $u^\dagger$ (left) and the ptimal state $y^\dagger$ and optimal adjoint state $p^\dagger$ (right) for Example 3.

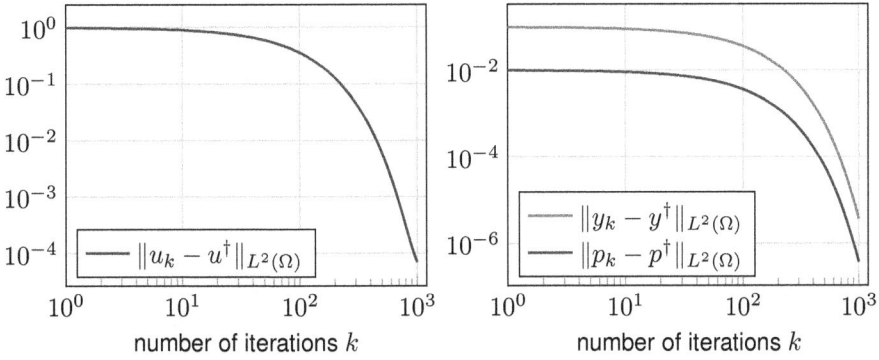

Figure 5.4: Regularization error for the control (left) and for the state and adjoint state (right) for Example 1.

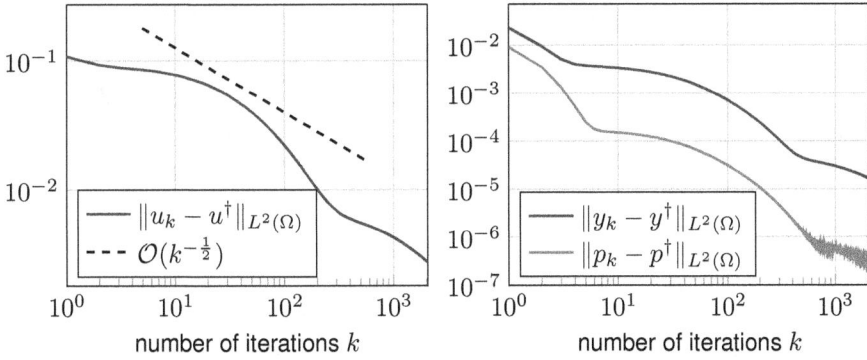

Figure 5.5: Regularization error for the control (left) and for the state and adjoint state (right) for Example 2.

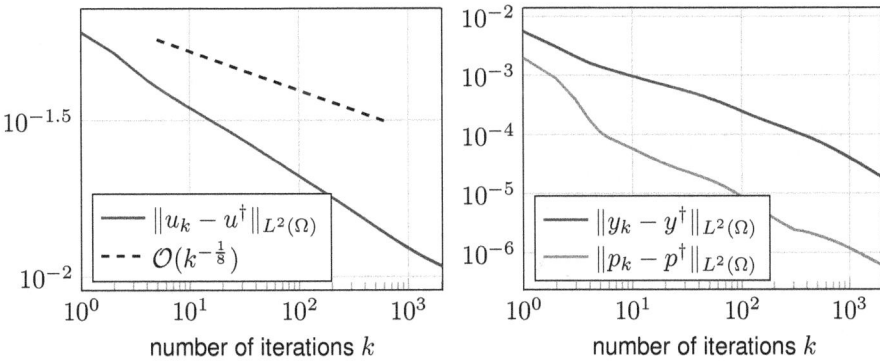

Figure 5.6: Regularization error for the control (left) and for the state and adjoint state (right) for Example 3.

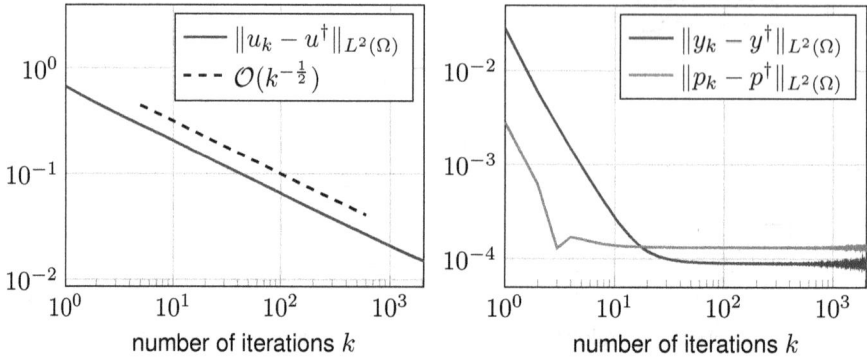

Figure 5.7: Regularization error for the control (left) and for the state and adjoint state (right) for Example 4.

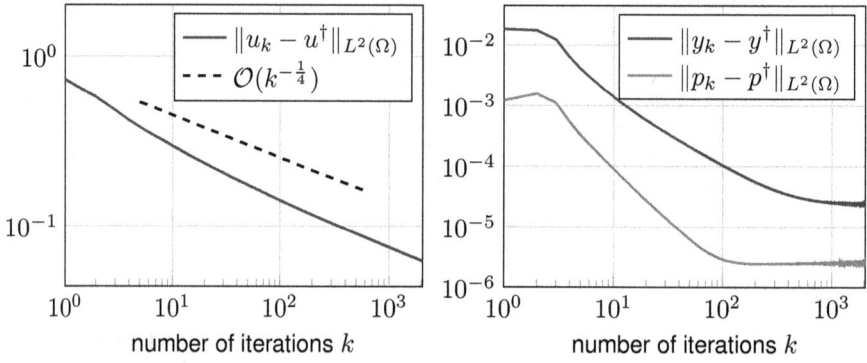

Figure 5.8: Regularization error for the control (left) and for the state and adjoint state (right) for Example 5.

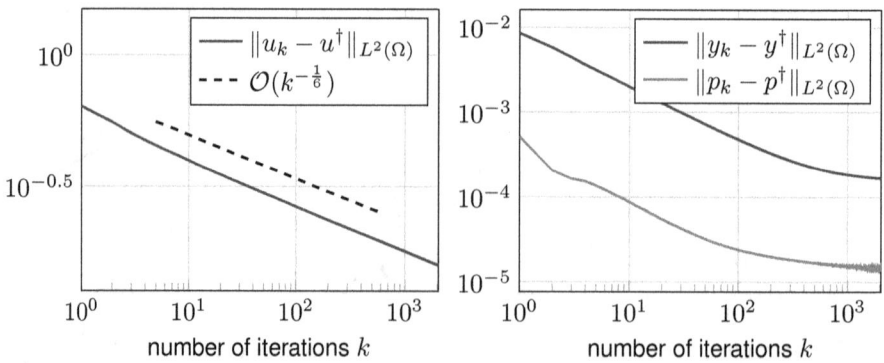

Figure 5.9: Regularization error for the control (left) and for the state and adjoint state (right) for Example 6.

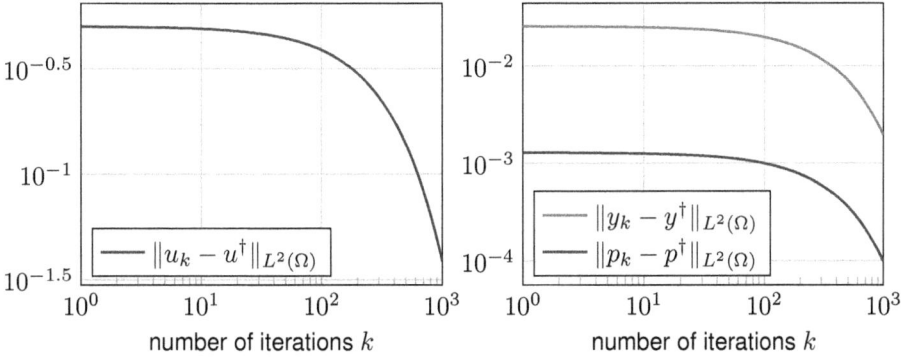

Figure 5.10: Regularization error for the control (left) and for the state and adjoint state (right) for Example 7.

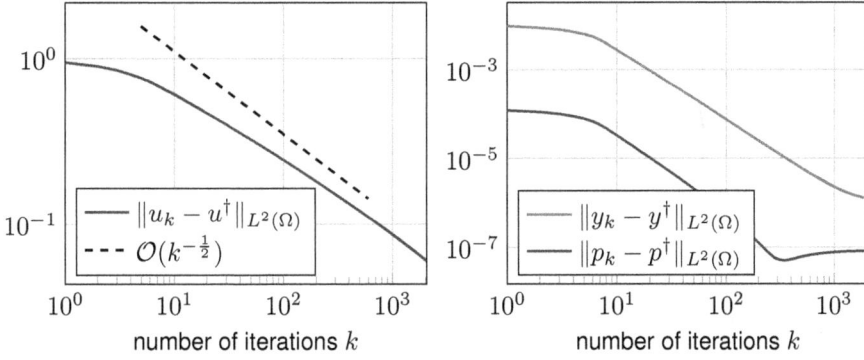

Figure 5.11: Regularization error for the control (left) and for the state and adjoint state (right) for Example 8.

CHAPTER 6

ALM for State Constraints and Sparsity

Until now, we have only considered linear quadratic optimal control problems with additional control constraints. However, in many physical systems it is needed to add additional state constraints. An example for such a situation is the optimal control of a heat source discussed in Section 5.1.1. There the state y is the temperature distribution in the domain Ω. Now assume that for technical reasons the temperature should not exceed a certain threshold ψ to avoid mechanical damage through over-heating. The optimal control $\bar{u}$ and its state $\bar{y}$ should therefore satisfy $\bar{y}(x) \leq \psi(x)$ for all $x \in \Omega$.

Note that for the elliptic partial differential equation discussed in Section 5.1.1 we obtained that the state is a continuous function. So let us assume $\psi \in C(\bar{\Omega})$, hence the inequality $\bar{y} \leq \psi$ is to be understood pointwise in the space $C(\bar{\Omega})$. We will explain later why we use the space $C(\bar{\Omega})$ instead of using the embedding $C(\bar{\Omega}) \to L^2(\Omega)$ and interpreting this inequality in the L^2-sense.

These additional constraints increase the complexity of the problem. In this chapter we develop a method to solve optimal control problem of the form

$$
\begin{aligned}
\text{Minimize} \quad & \frac{1}{2}\|y - z\|_{L^2(\Omega)}^2 + \beta\|u\|_{L^1(\Omega)} \\
\text{such that} \quad & Ay = u \ \text{ in } \ \Omega, \\
& y = 0 \ \text{ on } \ \partial\Omega, \\
& y \leq \psi \ \text{ in } \ \Omega, \\
& u_a \leq u \leq u_b \ \text{in} \ \Omega,
\end{aligned}
$$

with a given function $\psi \in C(\bar{\Omega})$ and $\beta \geq 0$. Here A is a linear elliptic operator. The details are made precise in the next section.

We apply an augmented Lagrange method (ALM), which was briefly introduced in Subsection 2.2.6. First we introduce the model problem in Section 6.1 and discuss the additional difficulties. In Section 6.2 and 6.3 we introduce the tools needed for the convergence analysis, e.g. optimality conditions. The augmented Lagrange method is then introduced in Section 6.4 and the convergence analysis is carried out in Section 6.5. The results of this chapter can be found in the publication [58].

6.1   Model Problem with State Constraints

Throughout this section we will use the same notation (P) for the optimal control as in Section 3.1. This will not cause any problems as the contents of the sections do not intersect. In this section we consider a convex optimal control problem of the following form

$$\min \ J(y, u) := \frac{1}{2}\|y - y_d\|^2_{L^2(\Omega)} + \beta\|u\|_{L^1(\Omega)}, \qquad (P)$$

subject to

$$Ay = u \quad \text{in } \Omega$$
$$y = 0 \quad \text{on } \partial\Omega$$
$$y \leq \psi \quad \text{in } \bar{\Omega}$$
$$u_a \leq u \leq u_b \quad \text{in } \Omega$$

and $u \in L^2(\Omega)$. We set $j(u) := \|u\|_{L^1(\Omega)}$ for abbreviation. Here A is a linear elliptic operator and $\beta \geq 0$. The main difficulties of this problem are the pointwise state constraints $y(x) \leq \psi(x)$ and the convex but non-differentiable term $\|u\|_{L^1(\Omega)}$. Note that there is no additional L^2-regularization term present in (P), which makes the problem ill-posed and numerically challenging.

The motivation for the L^1-term in the cost functional is the following. The optimal solution $\bar{u}$ of (P) is sparse, i.e., the control is zero on large parts of the domain if β is large enough. This can be used in the optimal placement of controllers, especially in situations where it is not desirable to control the system from the whole domain Ω, see [89]. Such sparsity promoting optimal control problems without state constraints have been studied in, e.g. [97–99] for optimal control of linear partial differential equations and in [13, 16] for the optimal control of semilinear equations. For sufficient second-order conditions for the state constrained sparsity promoting optimal control problem with a semilinear partial differential equation we refer to [18].

In order to deal with the state constraints we apply an augmented Lagrange method established by Karl and Wachsmuth in [59]. There the optimal control problem

$$\text{Minimize } \frac{1}{2}\|y - y_d\|^2_{L^2(\Omega)} + \frac{\alpha}{2}\|u\|^2_{L^2(\Omega)} \qquad (6.1)$$

with $\alpha > 0$ subject to an elliptic linear partial differential equation, state constraints and bilateral control constraints had been considered. Under suitable regularity assumptions the existence of Lagrange multipliers can be proven. However, in many cases the multiplier associated with the state constraint $\bar{\mu}$ has a very low regularity, e.g. $\bar{\mu} \in C(\Omega)^* = \mathcal{M}(\bar{\Omega})$, where $\mathcal{M}(\bar{\Omega})$ denotes the space of regular Borel measures on $\bar{\Omega}$. This makes the numerical solution of (P) very challenging. Although augmented Lagrange methods for inequality constraints are well known in finite dimensional spaces, only a few publications considering state constraints in infinite dimensional spaces are available: In [2,3] the state equation is augmented, and in [50] they deal with finitely many state constraints. Let us also mention the quite general approach for augmented Lagrange for optimization problems in Banach spaces which was recently established in [55].

Apart from the augmented Lagrange method there exist some other different approaches to deal with state constraints. We want to mention [66], in which a simultaneous Tikhonov and Lavrentiev regularization had been applied for (P). There the motivation was to derive error estimates under a source condition and the assumption that the state constraints are not active for solutions of (P). Furthermore they assumed that for the lower bound on the control it holds $u_a = 0$. In this paper we do not assume any of the above, which allows us to apply our method to a bigger class of problems.

Our aim is to modify and extend the method presented in [59] to obtain a numerical scheme to solve (P). The main idea is the following. We add a Tikhonov regularization term $\frac{\alpha}{2}\|u\|_{L^2(\Omega)}^2$ to (P) and apply the augmented Lagrange method. Thus, in every iteration we examine the optimal control problem

$$
\text{Minimize } \frac{1}{2}\|y - y_d\|_{L^2(\Omega)}^2 + \beta\|u\|_{L^1(\Omega)} + \frac{\alpha}{2}\|u\|_{L^2(\Omega)}^2
$$
$$
+ \frac{1}{2\rho}\int_\Omega \left((\mu + \rho(y - \psi))_+\right)^2 - \mu^2 \ \mathrm{d}x \tag{6.2}
$$

subject to an elliptic partial differential equation and bilateral control constraints. We denote $f_+(x) := \max(f(x), 0)$. Here, again $\alpha > 0$ denotes the regularization parameter of the Tikhonov term, while ρ is the penalization parameter of the augmented state constraints. Both variables are coupled in our method. During the algorithm we decrease the regularization parameter $\alpha \to 0$ while increasing the penalization parameter ρ. The coupling is described in detail in Section 6.4. Since the decrease of α is a classical Tikhonov regularization approach, we aim to achieve strong convergence towards the solution of (P). Under mild assumptions, i.e. there exists a feasible point, the problem (P) has a unique solution, see Theorem 6.2.4.

Denote $\bar{u}$ the solution of (P), u^α the solution of (6.1) and u_ρ^α the solution of (6.2). Similar to [66] we split the error into the *Tikhonov error* and the *Lagrange error* in order to show convergence of the algorithm

$$
\|\bar{u} - u_\rho^\alpha\|_{L^2(\Omega)} \leq \underbrace{\|\bar{u} - u^\alpha\|_{L^2(\Omega)}}_{\text{Tikhonov error}} + \underbrace{\|u^\alpha - u_\rho^\alpha\|_{L^2(\Omega)}}_{\text{Lagrange error}}.
$$

We start this chapter by introducing and analysing the model problem in Section 6.2. In Section 6.3 we analyse the Tikhonov error. The augmented Lagrange method is introduced in Section 6.4 and the associated convergence analysis is carried out in Section 6.5.

6.2 Preliminary Results

Before we introduce the augmented Lagrange method we want to recall some important definitions and results.

6.2.1 Problem Setting

Let $\Omega \subset \mathbb{R}^N$, $N = \{1, 2, 3\}$ be a bounded Lipschitz domain. Let Y denote the space $Y := H_0^1(\Omega) \cap C(\bar{\Omega})$ and $U := L^2(\Omega)$. We want to solve the following state constrained optimal control problem: Minimize

$$J(y, u) := \frac{1}{2}\|y - y_d\|_{L^2(\Omega)}^2 + \beta\|u\|_{L^1(\Omega)}$$

over all $(y, u) \in Y \times U$ subject to the linear elliptic equation

$$(Ay)(x) = u(x) \quad \text{in } \Omega,$$
$$y(x) = 0 \qquad \text{on } \partial\Omega,$$

and subject to the pointwise state and control constraints

$$y(x) \leq \psi(x) \quad \text{in } \Omega,$$
$$u_a(x) \leq u(x) \leq u_b(x) \quad \text{a.e. in } \Omega.$$

In the sequel, we will work with the following set of standing assumptions.

Assumption 6.2.1. *We assume the following assumptions.*

1. *The given data satisfy $y_d \in L^2(\Omega)$, $u_a, u_b \in L^\infty(\Omega)$ with $u_a < 0 < u_b$ and $\psi \in C(\bar{\Omega})$.*

2. *The differential operator A is given by*

$$(Ay)(x) := -\sum_{i,j=1}^N \partial_{x_j}(a_{ij}(x)\partial_{x_i}y(x)),$$

   with $a_{i,j} \in C^{0,1}(\bar{\Omega})$. The operator A is assumed to be uniformly elliptic, i.e., there is $\delta > 0$ such that

$$\sum_{i,j=1}^N a_{ij}(x)\xi_i\xi_j \geq \delta|\xi|^2, \quad \forall \xi \in \mathbb{R}^N, \text{ a.e. on } \Omega.$$

The following theorem is taken from [15, Theorem 2.1].

Theorem 6.2.2. *For every $u \in L^2(\Omega)$ there exists a unique weak solution $y \in H_0^1(\Omega) \cap C(\bar{\Omega})$ of the state equation and it holds*

$$\|y\|_{H_0^1(\Omega)} + \|y\|_{C(\bar{\Omega})} \leq c\|u\|_{L^2(\Omega)},$$

with a constant $c > 0$ independent of u.

With this assumption one can prove the following properties of the control-to-state mapping S.

Theorem 6.2.3. *The control-to-state operator $S : L^2(\Omega) \to H_0^1(\Omega) \cap C(\bar{\Omega}), u \mapsto y$ is a linear, continuous, and compact operator.*

Proof. The linearity follows directly by the definition of S and for the compactness we refer to [15, Theorem 2.1]. □

In the following, we will use the feasible sets with respect to the state and control constraints denoted by

$$U_{ad} = \{u \in L^\infty(\Omega) \mid u_a(x) \le u(x) \le u_b(x) \text{ a.e. in } \Omega\},$$
$$Y_{ad} = \{y \in C(\bar{\Omega}) \mid y(x) \le \psi(x) \, \forall x \in \Omega\}.$$

The feasible set of the optimal control problem is denoted by

$$F_{ad} = \{(y, u) \in Y \times U \mid (y, u) \in Y_{ad} \times U_{ad}, \, y = Su\}.$$

The assumption $u_a < 0 < u_b$ is not a restriction. Assume that $u_a \ge 0$ on a subset $\Omega_1 \subseteq \Omega$. Then we can decompose the L^1-norm for $u \in U_{ad}$ as $\|u\|_{L^1(\Omega)} = \|u\|_{L^1(\Omega \setminus \Omega_1)} + \int_{\Omega_1} u$. Hence, on Ω_1 the L^1-norm is a linear functional and its treatment does not impose any further difficulties.

Theorem 6.2.4. *Assume that the feasible set F_{ad} is non-empty. Then, there exists a unique solution $\bar{u}$ with associated state $\bar{y}$ of (P).*

Proof. The existence of solutions follows by standard arguments. Due to the assumptions the operator S is linear, continuous, and injective. Hence the problem (P) is convex leading to a unique optimal state $\bar{y}$. By using the injectivity of S we now obtain uniqueness of the optimal control. □

6.2.2 Subdifferential of the L^1-Norm

In this section we want to recall some basic properties of the subdifferential of the function $j(u) = \|u\|_{L^1(\Omega)}$. Since j is convex and Lipschitz, the generalized gradient [26] and the subdifferential in the sense of convex analysis coincide. Following [13, 18] we know that $\lambda \in \partial j(u) \subseteq L^\infty(\Omega)$ if and only if

$$\lambda \begin{cases} = +1 & \text{if } u(x) > 0, \\ = -1 & \text{if } u(x) < 0, \\ \in [-1, +1] & \text{if } u(x) = 0. \end{cases}$$

Since j is a convex function with $\text{dom}(j) = L^1(\Omega)$ the subdifferential is always nonempty. A straightforward calculation now reveals that for every $\lambda \in \partial j(u)$ it holds

$$\int_\Omega \lambda(v - u) \, dx \le \|v\|_{L^1(\Omega)} - \|u\|_{L^1(\Omega)} \quad \forall v \in L^1(\Omega).$$

For more information we refer to the book of Bonnans and Shapiro [8, Section 2.4.3]. We need the subdifferential to establish first order condition, as we need to compute derivatives of the objective functional $\frac{1}{2}\|y - y_d\|^2 + \beta\|u\|_{L^1(\Omega)}$.

6.2.3 Optimality Conditions

We now want to establish optimality conditions for problem (P). Due to the additional state constraints the existence of Lagrange multipliers cannot be guaranteed without any further regularity assumptions. Here we will now benefit from the formulation of $y \leq \psi$ in $C(\bar{\Omega})$. Throughout this chapter we will assume that the following Slater condition is satisfied.

Assumption 6.2.5. *We assume that there exists $\hat{u} \in U_{\mathrm{ad}}$ and $\sigma > 0$ such that for $\hat{y} := S\hat{u}$ it holds*
$$\hat{y}(x) \leq \psi(x) - \sigma \quad \forall x \in \Omega.$$

Let us motivate this assumption. The constraint $y \leq \psi$ in $C(\bar{\Omega})$ can be rewritten as
$$-(y - \psi) \in K_C := \{z \in C(\bar{\Omega}) : z(x) \geq 0 \quad \forall x \in \Omega\}.$$

It is straightforward to check that K_C is a convex cone. Following [93, Theorem 6.1] we obtain the existence of a Lagrange multiplier in the dual space of $C(\bar{\Omega})$ if there exists a $\tilde{y}$ such that $-(\tilde{y} - \psi) \in \mathrm{int}(K_C)$ with respect to the natural norm in $C(\bar{\Omega})$. It is clear that the existence of such an inner point is guaranteed by Assumption 6.2.5. One major drawback is that the multiplier lies in $C(\bar{\Omega})^* = \mathcal{M}(\bar{\Omega})$ which is the space of regular Borel measures on $\bar{\Omega}$.

If we interpret the inequality $y \leq \psi$ in the bigger space $L^2(\Omega)$ the associated set
$$K_L := \{z \in L^2(\Omega) : z(x) \geq 0 \quad \text{for almost all } x \in \Omega\}$$

has empty interior. Hence [93, Theorem 6.1] is not applicable here.

Moreover, since S is linear, Assumption 6.2.5 is equivalent to the linearized Slater condition, which on the other hand implies the more general Zowe-Kurcyusz regularity condition. However, one already has to know the solution of the optimal control problem (P) to check whether the Zowe-Kurcyusz condition is satisfied. This is not the case for the proposed Slater condition.

Theorem 6.2.6. *Let $(\bar{u}, \bar{y})$ be a solution of the problem (P). Furthermore, let Assumption 6.2.5 be fulfilled. Then, there exists an adjoint state $\bar{p} \in W_0^{1,s}(\Omega)$, $s \in [1, N/(N-1))$, a Lagrange multiplier $\bar{\mu} \in \mathcal{M}(\bar{\Omega})$ and a subgradient $\bar{\lambda} \in \partial j(\bar{u})$ such that the following optimality system*

$$\begin{cases} A\bar{y} = \bar{u} & \text{in } \Omega, \\ \bar{y} = 0 & \text{on } \partial\Omega, \end{cases} \tag{6.3a}$$

$$\begin{cases} A^*\bar{p} = \bar{y} - y_d + \bar{\mu} & \text{in } \Omega, \\ \bar{p} = 0 & \text{on } \partial\Omega, \end{cases} \tag{6.3b}$$

$$(\bar{p} + \beta\bar{\lambda}, u - \bar{u})_{L^2(\Omega)} \geq 0 \quad \forall u \in U_{\mathrm{ad}}, \tag{6.3c}$$

$$\langle \bar{\mu}, \bar{y} - \psi \rangle_{\mathcal{M}(\bar{\Omega}),C(\bar{\Omega})} = 0, \quad \bar{\mu} \geq 0, \tag{6.3d}$$

is fulfilled. Here, the inequality $\bar{\mu} \geq 0$ means that $\langle \bar{\mu}, \varphi \rangle_{\mathcal{M}(\bar{\Omega}),C(\bar{\Omega})} \geq 0$ holds for all $\varphi \in C(\bar{\Omega})$ with $\varphi \geq 0$.

Proof. The proof can be found in [18, Theorem 2.5]. $\qquad\square$

In the definition (6.3b) for the optimal adjoint state $\bar{p}$ we have to solve an elliptic equation with a measure on the right hand side. This problem is well-posed in the following sense.

Theorem 6.2.7. *Let $\bar{\mu} \in \mathcal{M}(\bar{\Omega})$ be a regular Borel measure. Then the adjoint state equation*

$$A^*\bar{p} = \bar{y} - y_d + \bar{\mu} \quad \text{in } \Omega,$$
$$\bar{p} = 0 \qquad\qquad \text{on } \partial\Omega$$

has a unique very weak solution $\bar{p} \in W_0^{1,s}(\Omega)$ with $s \in [1, N/(N-1))$, and it holds

$$\|\bar{p}\|_{W_0^{1,s}(\Omega)} \leq c \left(\|\bar{y}\|_{L^2(\Omega)} + \|y_d\|_{L^2(\Omega)} + \|\bar{\mu}\|_{\mathcal{M}(\bar{\Omega})} \right).$$

Proof. This result is due to [12, Theorem 4]. $\qquad\square$

Here a very weak solution is to be understood in the sense of transposition [14, Section 2], i.e. $\bar{p} \in W_0^{1,s}(\Omega)$ satisfies

$$\int_\Omega \bar{p}A\varphi \, dx = \int_\Omega (\bar{y} - y_d)\varphi \, dx + \int_\Omega \varphi \, d\bar{\mu}, \quad \forall \varphi \in H^2(\Omega) \cap H_0^1(\Omega).$$

The next theorem shows the relation between the adjoint state and the control. One can see, that if β is large, the control will be zero on large parts of Ω. Hence $\bar{u}$ is sparse.

Lemma 6.2.8. *Let $\bar{u}, \bar{p}, \bar{\lambda}, \bar{\mu}$ satisfy the optimality system. (6.3a)-(6.3d). Then the following relations hold for $\theta > 0$:*

$$\bar{u}(x) \begin{cases} = u_a(x) & \text{if } \bar{p}(x) > \beta, \\ = u_b(x) & \text{if } \bar{p}(x) < -\beta, \\ = 0 & \text{if } |p(x)| < \beta, \\ \in [u_a(x), u_b(x)] & \text{if } |p(x)| = \beta, \end{cases}$$

$$\bar{\lambda}(x) = P_{[-1,+1]}\left(-\frac{1}{\beta}\bar{p}(x) \right),$$

$$\bar{u}(x) = P_{[u_a(x),u_b(x)]}\left(\bar{u}(x) - \theta(\bar{p}(x) + \beta\bar{\lambda}(x)) \right).$$

From the second formula it follows that $\bar{\lambda}$ is unique if the multiplier $\bar{\mu}$ and adjoint state $\bar{p}$ are unique.

Proof. The proof only uses the optimality (6.3c) and can be found in [13, Theorem 3.1]. $\qquad\square$

6.3 Convergence Analysis of the Regularized Problem

Solving the problem (P) directly is challenging for mainly two reasons. First, since the multiplier corresponding to the state constraints appears in form of a measure, it is not clear how to deal with the state constraints. For the control constraints many powerful methods are available. Here, we only want to mention the semi-smooth Newton solvers [46, 47] and the active-set methods [4]. However, it is not clear how to implement the state constraints into a direct solver. In [5, 51] active-set methods have been used to solve problems where the state constraints have been treated by a Moreau-Yosida regularization. In [51, 52] also relations between semi-smooth Newton methods and active-set methods have been established that can be used to prove fast local convergence. In this work we want to adapt the approach of a modified augmented Lagrange method that has been proposed in [59] to overcome the lack of the multiplier's regularity.

The second challenge is the ill-posedness of the original problem (P). There small perturbations of the given data y_d may lead to large errors in the associated optimal controls. To deal with this issue we will use the well-known Tikhonov regularization technique with some positive regularization parameter $\alpha > 0$. The regularized problem is given by

$$\text{Minimize } J_\alpha(y, u) := \frac{1}{2}\|y - y_d\|_{L^2(\Omega)}^2 + \beta\|u\|_{L^1(\Omega)} + \frac{\alpha}{2}\|u\|_{L^2(\Omega)}^2$$

$$\text{s.t.} \qquad Ay = u \quad \text{in } \Omega,$$
$$y = 0 \quad \text{on } \partial\Omega, \qquad\qquad\qquad (P^\alpha)$$
$$y \leq \psi,$$
$$u \in U_{\text{ad}}.$$

It is clear that (P^α) admits a unique solution u^α with associated state y^α. One can expect that u^α converges to the solution of (P) as $\alpha \to 0$. Similar results can be found in the literature, e.g. [97].

Theorem 6.3.1. *Let u^α be the unique solution of (P^α) with $\alpha > 0$ and associated state y^α. Furthermore let $\bar{u}$ be the unique solution of (P) and $\bar{y}$ its associated optimal state. Then we have for $\alpha \to 0$*

$$\|u^\alpha - \bar{u}\|_{L^2(\Omega)} \to 0,$$

$$\frac{1}{\alpha}\|y^\alpha - \bar{y}\|_{L^2(\Omega)}^2 \to 0.$$

Proof. We first show that $\|u^\alpha\|_{L^2(\Omega)} \leq \|\bar{u}\|_{L^2(\Omega)}$ for all $\alpha > 0$. To shorten our notation we set $y = Su$ in the cost functional of (P^α) and define the reduced cost functional $J_\alpha(u) := J_\alpha(Su, u)$. Let J_0 denote the cost functional J_α for $\alpha := 0$. We start with

$$J_0(u^\alpha) + \frac{\alpha}{2}\|u^\alpha\|_{L^2(\Omega)}^2 = J_\alpha(u^\alpha) \leq J_\alpha(\bar{u}) = J_0(\bar{u}) + \frac{\alpha}{2}\|\bar{u}\|_{L^2(\Omega)}^2 \leq J_0(u^\alpha) + \frac{\alpha}{2}\|\bar{u}\|_{L^2(\Omega)}^2,$$

where we exploited the optimality of u^α for (P^α) and the optimality of $\bar{u}$ for (P). This yields $\|u^\alpha\|_{L^2(\Omega)} \leq \|\bar{u}\|_{L^2(\Omega)}$. Now we use that the set U_{ad} is weakly compact and extract a subsequence $u^{\alpha_i} \rightharpoonup u^* \in U_{\mathrm{ad}}$. Since the operator S is compact, see Theorem 6.2.3, we obtain strong convergence of the associated states $y^{\alpha_i} \to y^* = Su^*$ in $H_0^1(\Omega) \cap C(\bar{\Omega})$. Now let $u \in U_{\mathrm{ad}}$ be arbitrary, then

$$J_0(u^*) = \lim_{i \to \infty} J_0(u^{\alpha_i}) = \lim_{i \to \infty} J_{\alpha_i}(u^{\alpha_i}) \le \lim_{i \to \infty} J_{\alpha_i}(u) = J_0(u).$$

Hence u^* is a minimizer of J_0. The solution $\bar{u}$ of (P) is unique and since the problems (P) and (P^α) coincide for $\alpha = 0$ we obtain $\bar{u} = u^*$. As the norm is weakly lower semicontinuous we get

$$\limsup_{i \to \infty} \|u^{\alpha_i}\|_{L^2(\Omega)} \le \|u^*\|_{L^2(\Omega)} \le \liminf_{i \to \infty} \|u^{\alpha_i}\|_{L^2(\Omega)} \le \limsup_{i \to \infty} \|u^{\alpha_i}\|_{L^2(\Omega)},$$

which shows $\|u^{\alpha_i}\|_{L^2(\Omega)} \to \|u^*\|_{L^2(\Omega)}$. As a well known fact, weak and norm convergence yield strong convergence and hence we have $u^{\alpha_i} \to u^*$. As the sequence u^{α_i} was arbitrarily chosen we obtain convergence of the whole sequence $u^\alpha \to \bar{u}$.
We now want to show improved convergence results for the states. Since the function

$$J : L^2(\Omega) \to \mathbb{R}, \quad y \mapsto \frac{1}{2}\|y - y_d\|_{L^2(\Omega)}^2$$

is a strongly convex function we know that the following inequality holds for all $t \in [0,1]$ and $y_1, y_2 \in L^2(\Omega)$ with some $m > 0$ (in fact $m = 2$)

$$J(ty_1 + (1-t)y_2) \le tJ(y_1) + (1-t)J(y_2) - \frac{1}{2}m \cdot t(1-t)\|y_1 - y_2\|_{L^2(\Omega)}^2.$$

Now let $(u, y) \in F_{\mathrm{ad}}$ and define $t = \frac{1}{2}$, $y_1 = \bar{y}$ and $y_2 = y$. Furthermore note, that with $(u, y), (\bar{u}, \bar{y}) \in F_{\mathrm{ad}}$ the convex combination is also feasible. To be precise we obtain with the optimality of $\bar{u}$ that

$$J(\bar{y}) \le J\left(\frac{1}{2}\bar{y} + \frac{1}{2}y\right) \le \frac{1}{2}J(\bar{y}) + \frac{1}{2}J(y) - \frac{m}{4}\|\bar{y} - y\|_{L^2(\Omega)}^2.$$

Rearranging this inequality above yields the following growth condition

$$J_0(\bar{u}) + c\|\bar{y} - y\|_{L^2(\Omega)}^2 \le J_0(u) \quad \forall (u, y) \in F_{\mathrm{ad}}.$$

Note that $J_0(\bar{u}) = J(\bar{y})$, since $\bar{y} = S\bar{u}$. This growth condition can now be used to establish improved convergence results for the states (y^α). Recall that $J_\alpha(u^\alpha) \le J_\alpha(\bar{u})$ and estimate

$$J_0(\bar{u}) + c\|y^\alpha - \bar{y}\|_{L^2(\Omega)}^2 + \frac{\alpha}{2}\|u^\alpha\|_{L^2(\Omega)}^2 \le J_0(u^\alpha) + \frac{\alpha}{2}\|u^\alpha\|_{L^2(\Omega)}^2 = J_\alpha(u^\alpha)$$

$$\le J_\alpha(\bar{u}) = J_0(\bar{u}) + \frac{\alpha}{2}\|\bar{u}\|_{L^2(\Omega)}^2.$$

This implies

$$\|y^\alpha - \bar{y}\|_{L^2(\Omega)}^2 \le c \cdot \alpha \left(\|\bar{u}\|_{L^2(\Omega)}^2 - \|u^\alpha\|_{L^2(\Omega)}^2\right).$$

Using the already established strong convergence $u^\alpha \to \bar{u}$, we get

$$\lim_{\alpha \to 0} \frac{1}{\alpha} \|y^\alpha - \bar{y}\|_{L^2(\Omega)}^2 = 0,$$

which finishes the proof.

$\square$

Let us assume that the Slater condition given in Assumption 6.2.5 is satisfied. Then first order necessary optimality conditions can be established for the regularized problem.

Theorem 6.3.2. *Let (u^α, y^α) be the solution of the problem (P^α). Furthermore, let Assumption 6.2.5 be fulfilled. Then, there exists an adjoint state $p^\alpha \in W^{1,s}(\Omega)$, $s \in [1, N/(N-1))$, a Lagrange multiplier $\mu^\alpha \in \mathcal{M}(\bar{\Omega})$ and a subgradient $\lambda^\alpha \in \partial j(u^\alpha)$ such that the following optimality system holds:*

$$\begin{cases} Ay^\alpha = u^\alpha & \text{in } \Omega, \\ y^\alpha = 0 & \text{on } \partial\Omega, \end{cases} \tag{6.4a}$$

$$\begin{cases} A^* p^\alpha = y^\alpha - y_d + \mu^\alpha & \text{in } \Omega, \\ p^\alpha = 0 & \text{on } \partial\Omega, \end{cases} \tag{6.4b}$$

$$(p^\alpha + \alpha u^\alpha + \beta\lambda^\alpha, u - u^\alpha)_{L^2(\Omega)} \geq 0 \quad \forall u \in U_{\text{ad}}, \tag{6.4c}$$

$$\langle \mu^\alpha, y^\alpha - \psi \rangle_{\mathcal{M}(\bar{\Omega}),C(\bar{\Omega})} = 0, \quad \mu^\alpha \geq 0. \tag{6.4d}$$

Proof. The proof can be found in [18, Theorem 2.5]. $\square$

Similar to Lemma 6.2.8 it is possible to reconstruct the optimal solution u^α by its associated adjoint state p^α on certain sets. The relation between u^α and p^α presented in the next lemma can be used in numerical algorithms, see e.g. the active-set method described in Chapter 7.

Lemma 6.3.3. *Let $u^\alpha, y^\alpha, p^\alpha, \lambda^\alpha, \mu^\alpha$ satisfy the optimality system (6.4). Then the following relations hold:*

$$u^\alpha(x) = \begin{cases} u_a(x) & \text{if } \beta - \alpha u_a(x) < p^\alpha(x), \\ \frac{1}{\alpha}(\beta - p^\alpha(x)) & \text{if } \beta \leq p^\alpha(x) \leq \beta - \alpha u_a(x), \\ 0 & \text{if } |p^\alpha(x)| < \beta, \\ \frac{1}{\alpha}(-\beta - p^\alpha(x)) & \text{if } -\alpha u_b(x) - \beta \leq p^\alpha(x) \leq -\beta, \\ u_b(x) & \text{if } p^\alpha(x) < -\alpha u_b(x) - \beta, \end{cases}$$

$$\lambda^\alpha(x) = P_{[-1,1]}\left(-\frac{1}{\beta}(p^\alpha(x) + \alpha u^\alpha(x))\right),$$

$$u^\alpha(x) = P_{[u_a(x),u_b(x)]}\left(-\frac{1}{\alpha}(p^\alpha(x) + \beta\lambda^\alpha(x))\right).$$

Proof. First note that by the variational inequality (6.4c) can be interpreted pointwise for almost all $x \in \Omega$, see [93, Lemma 2.26]

$$(p^\alpha(x) + \alpha u^\alpha(x) + \beta \lambda^\alpha(x))(u - u^\alpha(x)) \geq 0 \quad \forall u \in [u_a(x), u_b(x)]. \tag{6.5}$$

Consider the first case and assume $\beta - \alpha u_a(x) < p^\alpha(x)$. Note that $\lambda^\alpha \in \partial \|u^\alpha\|_{L^1(\Omega)}$, hence $\lambda^\alpha(x) \in [-1, 1]$. Now we obtain using $u_a(x) \leq u^\alpha(x)$

$$p^\alpha(x) + \alpha u^\alpha(x) + \beta \lambda^\alpha(x) \geq p^\alpha(x) + \alpha u_a(x) - \beta > 0.$$

Now (6.5) yields

$$u - u^\alpha(x) \geq 0 \quad \forall u \in [u_a(x), u_b(x)]$$

and we conclude $u^\alpha(x) = u_a(x)$. By our assumption we know $u_a(x) < 0$, hence

$$-1 = \lambda^\alpha(x) = P_{[-1,1]}\left(-\frac{1}{\beta}(p^\alpha(x) + \alpha u^\alpha(x))\right).$$

With a similar argument we handle the case $p^\alpha(x) < -\alpha u_b(x) - \beta$.
Now assume that $\beta \leq p^\alpha(x) \leq \beta - \alpha u_a(x)$ holds. We prove $u^\alpha(x) = \frac{1}{\alpha}(\beta - p^\alpha(x))$ by contradiction. First assume $u^\alpha(x) < \frac{1}{\alpha}(\beta - p^\alpha(x))$. This leads to $u^\alpha(x) < 0$ hence $\lambda^\alpha(x) = -1$. We now compute

$$p^\alpha(x) + \alpha u^\alpha(x) + \beta \lambda^\alpha(x) < p^\alpha(x) + \beta - p^\alpha(x) - \beta = 0.$$

From (6.5) we conclude $u^\alpha(x) = u_b(x) > 0$ which is a contradiction. Now assume $u^\alpha(x) > \frac{1}{\alpha}(\beta - p^\alpha(x))$. From this inequality we obtain $\alpha u^\alpha(x) > \beta - p^\alpha(x) \geq \alpha u_a(x)$ hence $u^\alpha(x) > u_a(x)$. We now obtain using $\lambda^\alpha(x) \in [-1, 1]$

$$p^\alpha(x) + \alpha u^\alpha(x) + \beta \lambda^\alpha(x) > p^\alpha(x) + \beta - p^\alpha(x) - \beta = 0$$

and we obtain $u^\alpha(x) = u_a(x)$ which again is a contradiction. A straightforward calculation now reveals

$$\lambda^\alpha(x) = P_{[-1,1]}\left(-\frac{1}{\beta}(p^\alpha(x) + \alpha u^\alpha(x))\right).$$

The case $-\alpha u_b(x) - \beta \leq p^\alpha(x) \leq -\beta$ is handled with a similar calculation.
We now consider the remaining case $|p^\alpha(x)| < \beta$. Let us show that $u^\alpha(x) = 0$ holds by contradiction. Assume $u^\alpha(x) > 0$, hence $\lambda^\alpha(x) = 1$ and

$$p^\alpha(x) + \alpha u^\alpha(x) + \beta \lambda^\alpha(x) > p^\alpha(x) + \beta > 0$$

leading to $u^\alpha(x) = u_a(x) < 0$ which is a contradiction. With a similar argument one can handle the case $u^\alpha(x) < 0$. With $u^\alpha(x) = 0$ and $u_a(x) < 0 < u_b(x)$ we conclude from (6.5) that

$$\lambda^\alpha(x) = -\frac{1}{\beta}p^\alpha(x) = P_{[-1,1]}\left(-\frac{1}{\beta}(p^\alpha(x) + \alpha u^\alpha(x))\right)$$

holds. $\qquad\square$

In the subsequent analysis we will need that the multipliers for the problem (P^α) are uniformly bounded for all $\alpha \geq 0$. Note that for $\alpha = 0$ the problem (P^α) reduces to problem (P). The boundedness of the multipliers can be expected from abstract theory [8] and [93, Theorem 6.3], and we make use of the Slater condition to prove it.

Lemma 6.3.4. *Let $0 \leq \alpha \leq C$ and define the set*

$$M^\alpha := \{\mu^\alpha \in \mathcal{M}(\bar{\Omega}) : (u^\alpha, y^\alpha, p^\alpha, \lambda^\alpha, \mu^\alpha) \text{ satisfy } (6.4a) - (6.4d)\}$$

of all multipliers associated with problem (P^α). Then the multipliers are uniformly bounded, i.e. there exists a constant $C > 0$ independent from α such that

$$\|\mu^\alpha\|_{\mathcal{M}(\bar{\Omega})} \leq C, \quad \forall \alpha \geq 0 \ \forall \mu^\alpha \in M^\alpha.$$

Proof. We follow the book of Tröltzsch [93] and consider our solution mapping $S : L^2(\Omega) \to C(\bar{\Omega})$. Then the dual operator is a mapping $S^* : \mathcal{M}(\bar{\Omega}) \to L^2(\Omega)$. Let $\alpha \geq 0$ be given, and u^α, y^α be the solution of (P^α) with an associated multiplier μ^α. We now use the Slater condition from Assumption 6.2.5 and compute for any $f \in C(\bar{\Omega})$ with $\|f\|_\infty = 1$

$$\sigma \left| \int_\Omega f \, d\mu^\alpha \right| \leq \sigma \int_\Omega |f| \, d\mu^\alpha \leq \int_\Omega \sigma \, d\mu^\alpha \leq \int_\Omega (\psi - \hat{y}) \, d\mu^\alpha$$

$$= \underbrace{\langle \mu^\alpha, \psi - y^\alpha \rangle_{\mathcal{M}(\bar{\Omega}), C(\bar{\Omega})}}_{=0 \text{ by } (6.4d)} + \langle \mu^\alpha, y^\alpha - \hat{y} \rangle_{\mathcal{M}(\bar{\Omega}), C(\bar{\Omega})}$$

$$= \langle \mu^\alpha, S(u^\alpha - \hat{u}) \rangle_{\mathcal{M}(\bar{\Omega}), C(\bar{\Omega})}$$

$$= \int_\Omega (S^* \mu^\alpha)(u^\alpha - \hat{u}) \, dx.$$

Now recall that the adjoint equation (6.4b) can be rewritten as

$$S^* \mu^\alpha = S^*(y_d - S u^\alpha) + p^\alpha.$$

Furthermore by the assumption $u^\alpha \in U_{ad}$ and by Theorem 6.2.4 and 6.2.7 we obtain that u^α and y^α are uniformly bounded in $L^2(\Omega)$. We now get

$$\sigma \|\mu^\alpha\|_{\mathcal{M}(\bar{\Omega})} = \sigma \sup_{f \in C(\bar{\Omega}), \, \|f\|_\infty = 1} \left| \int_\Omega f \, d\mu^\alpha \right| \leq \int_\Omega (S^* \mu^\alpha)(u^\alpha - \hat{u}) \, dx,$$

leading to

$$\sigma \|\mu^\alpha\|_{\mathcal{M}(\bar{\Omega})} \leq \int_\Omega (S^*(y_d - S u^\alpha) + p^\alpha)(u^\alpha - \hat{u}) \, dx$$

$$= \int_\Omega (S^*(y_d - S u^\alpha))(u^\alpha - \hat{u}) \, dx + \int_\Omega p^\alpha (u^\alpha - \hat{u}) \, dx.$$

We now apply the optimality condition (6.4c) and obtain with the boundedness of λ^α, see Lemma 6.3.3 and the boundedness of α that the following holds

$$\sigma\|\mu^\alpha\|_{\mathcal{M}(\bar\Omega)} \le c\|u^\alpha - \hat u\|_{L^2(\Omega)}\|y_d - y^\alpha\|_{L^2(\Omega)} + \int_\Omega (\alpha u^\alpha + \beta\lambda^\alpha)(\hat u - u^\alpha)\,\mathrm{d}x$$

$$\le c\|u^\alpha - \hat u\|_{L^2(\Omega)}\left(\|y_d - y^\alpha\|_{L^2(\Omega)} + \|\alpha u^\alpha + \beta\lambda^\alpha\|_{L^2(\Omega)}\right)$$

$$\le c.$$

Dividing the above inequality by $\sigma > 0$ finishes the proof. $\qquad\square$

6.4 The Augmented Lagrange Method

In the following we want to solve the regularized problem (P^α) for $\alpha \to 0$. For fixed α we follow the idea presented in [59] and replace the inequality constraint $y \le \psi$ by an augmented penalization term. In that way we get rid of the measure and work instead with an approximation.

6.4.1 The Augmented Lagrange Optimal Control Problem

First let us introduce the penalty function P which we use to augment the state constraints. Let $\rho > 0$ be a given penalty parameter, and let $\mu \in L^2(\Omega)$ with $\mu \ge 0$ be a given approximation of the Lagrange multiplier. Now we define

$$P(y,\rho,\mu) := \frac{1}{2\rho}\int_\Omega \left((\mu + \rho(y-\psi))_+\right)^2 - \mu^2 \,\mathrm{d}x.$$

Here $f_+(x) := \max(f(x), 0)$. Let now $\rho > 0$ and $\mu \in L^2(\Omega)$ be given. Then in each step of the augmented Lagrange method the following subproblem has to be solved: Minimize

$$J_\rho^\alpha(y,u,\mu) := \frac{1}{2}\|y - y_d\|_{L^2(\Omega)}^2 + \beta\|u\|_{L^1(\Omega)} + \frac{\alpha}{2}\|u\|_{L^2(\Omega)}^2 + P(y,\rho,\mu) \qquad (P_{\alpha,\rho,\mu})$$

with $\alpha > 0$, subject to the state equation and the control constraints

$$y = Su, \quad u \in U_{\mathrm{ad}}.$$

A solution of $(P_{\alpha,\rho,\mu})$ will be denoted by u_ρ^α with associated state y_ρ^α and adjoint state p_ρ^α. The next theorem shows that the subproblem is uniquely solveable.

Theorem 6.4.1. *For every $\rho > 0$, $\mu \in L^2(\Omega)$ with $\mu \ge 0$ the augmented Lagrange control problem $(P_{\alpha,\rho,\mu})$ admits a unique solution $u_\rho^\alpha \in U_{\mathrm{ad}}$ with associated optimal state $y_\rho^\alpha \in Y$ and adjoint state p_ρ^α.*

Proof. Since U_{ad} is closed, bounded and convex and J_ρ^α is coercive, weakly lower semi-continuous and strictly convex, problem $(P_{\alpha,\rho,\mu})$ has a unique solution $u_\rho^\alpha \in U_{\mathrm{ad}}$. For more details see [93] and [27]. $\qquad\square$

Solutions of $(P_{\alpha,\rho,\mu})$ can be characterized by the first-order optimality conditions.

Theorem 6.4.2 (First-order necessary optimality conditions). *Let $(u_\rho^\alpha, y_\rho^\alpha)$ be the solution of $(P_{\alpha,\rho,\mu})$. Then, there exists a unique adjoint state $p_\rho^\alpha \in H_0^1(\Omega)$ associated with the optimal control u_ρ^α and a subdifferential $\lambda_\rho^\alpha \in \partial j(u_\rho^\alpha)$, satisfying the following system.*

$$\begin{cases} Ay_\rho^\alpha = u_\rho^\alpha & \text{in } \Omega, \\ y_\rho^\alpha = 0 & \text{on } \partial\Omega, \end{cases} \tag{6.6a}$$

$$\begin{cases} A^*p_\rho^\alpha = y_\rho^\alpha - y_d + \mu_\rho^\alpha & \text{in } \Omega, \\ p_\rho^\alpha = 0 & \text{on } \partial\Omega, \end{cases} \tag{6.6b}$$

$$(p_\rho^\alpha + \alpha u_\rho^\alpha + \beta\lambda_\rho^\alpha, u - u_\rho^\alpha)_{L^2(\Omega)} \geq 0 \quad \forall u \in U_{\text{ad}}, \tag{6.6c}$$

$$\mu_\rho^\alpha := \left(\mu + \rho(y_\rho^\alpha - \psi)\right)_+. \tag{6.6d}$$

Proof. Can be proven by extending the results in [48, Corollary 1.3, p. 73]. □

Note that we can extend the results obtained in Lemma 6.3.3 to the solution of $(P_{\alpha,\rho,\mu})$. The proof only uses the variational inequality (6.4c) and can therefore directly be copied to (6.6c).

Further we make an analogous observation like in [59]. Boundedness of μ_ρ^α in the L^1-norm is enough to get boundedness of the solution $(u_\rho^\alpha, y_\rho^\alpha, p_\rho^\alpha, \lambda_\rho^\alpha)$ of (6.6).

Theorem 6.4.3. *Let $\rho > 0$ and $\mu \in L^2(\Omega)$ be given. Let $s \in [1, N/(N-1))$ and α be bounded. Then there exists a constant $c > 0$ independent of $\alpha, \rho,$ and μ such that for all solutions $(u_\rho^\alpha, y_\rho^\alpha, p_\rho^\alpha, \mu_\rho^\alpha)$ of (6.6) it holds*

$$\|y_\rho^\alpha\|_{H^1(\Omega)} + \|y_\rho^\alpha\|_{C(\bar\Omega)} + \|u_\rho^\alpha\|_{L^2(\Omega)} + \|p_\rho^\alpha\|_{W^{1,s}(\Omega)} \leq c(\|\mu_\rho^\alpha\|_{L^1(\Omega)} + 1).$$

Proof. The proof just differs from the one of [59, Theorem 3.3] concerning the additional subgradient in (6.6b). Hence, we just give the most important steps here. Let us test the state equation (6.6a) with p_ρ^α and the adjoint equation (6.6b) with y_ρ^α. This yields

$$(p_\rho^\alpha, u_\rho^\alpha)_{L^2(\Omega)} = (y_\rho^\alpha - y_d, y_\rho^\alpha)_{L^2(\Omega)} + (\mu_\rho^\alpha, y_\rho^\alpha)_{L^2(\Omega)}.$$

Now fix a $u \in U_{\text{ad}}$ and use it in (6.6c), yielding

$$(y_\rho^\alpha - y_d, y_\rho^\alpha)_{L^2(\Omega)} + (\mu_\rho^\alpha, y_\rho^\alpha)_{L^2(\Omega)}$$
$$\leq (\alpha u_\rho^\alpha, u - u_\rho^\alpha)_{L^2(\Omega)} + (p_\rho^\alpha, u)_{L^2(\Omega)} + (\beta\lambda_\rho^\alpha, u - u_\rho^\alpha)_{L^2(\Omega)}.$$

By Young's inequality and exploiting $(\lambda_\rho^\alpha, u - u_\rho^\alpha)_{L^2(\Omega)} \leq \|u\|_{L^1(\Omega)} - \|u_\rho^\alpha\|_{L^1(\Omega)}$, we have

$$\frac{1}{2}\|y_\rho^\alpha\|_{L^2(\Omega)}^2 + \frac{\alpha}{2}\|u_\rho^\alpha\|_{L^2(\Omega)}^2 \leq \frac{1}{2}\|y_d\|_{L^2(\Omega)}^2 + \|\mu_\rho^\alpha\|_{L^1(\Omega)}\|y_\rho^\alpha\|_{C(\bar\Omega)} + \frac{\alpha}{2}\|u\|_{L^2(\Omega)}^2$$
$$+ \|p_\rho^\alpha\|_{L^2(\Omega)}\|u\|_{L^2(\Omega)} + \beta\left(\|u\|_{L^1(\Omega)} - \|u_\rho^\alpha\|_{L^1(\Omega)}\right).$$

Let us fix $\bar{s} \in (1, N/(N-1))$ such that $W^{1,\bar{s}}(\Omega)$ is continuously embedded in $L^2(\Omega)$. From Theorem 6.2.2 we now get $\|y_\rho^\alpha\|_{H^1(\Omega)} + \|y_\rho^\alpha\|_{C(\bar{\Omega})} \leq c\|u_\rho^\alpha\|_{L^2(\Omega)}$ and from Theorem 6.2.7 we get $\|p_\rho^\alpha\|_{L^2(\Omega)} \leq c\left(\|y_\rho^\alpha\|_{L^2(\Omega)} + \|y_d\|_{L^2(\Omega)} + \|\mu_\rho^\alpha\|_{L^1(\Omega)}\right)$. Now using the fact that u_ρ^α is bounded in $L^2(\Omega)$ and u is fixed to obtain the result.

$\square$

6.4.2 The Prototypical Augmented Lagrange Algorithm

In the following, let $(P_{\alpha,\rho,\mu}^k)$ denote the augmented Lagrange subproblem $(P_{\alpha,\rho,\mu})$ for given penalty parameter $\rho := \rho_k$, multiplier $\mu := \mu_k$ and regularization parameter $\alpha := \alpha_k$. We will denote its solution by $(\bar{u}_k, \bar{y}_k)$ with adjoint state $\bar{p}_k$ and updated multiplier $\bar{\mu}_k$, which is given by (6.6d).

Algorithm 6.5. *Let $\alpha_1 > 0$, $\rho_1 > 0$ and $\mu_1 \in L^2(\Omega)$ be given with $\mu_1 \geq 0$. Choose $\theta > 1$.*

1. *Solve $(P_{\alpha,\rho,\mu}^k)$ and obtain $(\bar{u}_k, \bar{y}_k, \bar{p}_k, \bar{\lambda}_k)$.*

2. *Set $\bar{\mu}_k := (\mu_k + \rho_k(\bar{y}_k - \psi))_+$.*

3. *If the step is successful set $\mu_{k+1} := \bar{\mu}_k$, $\rho_{k+1} := \rho_k$ and choose $0 < \alpha_{k+1}$ such that $\alpha_{k+1} < \alpha_k$.*

4. *Otherwise set $\mu_{k+1} := \bar{\mu}_k$ and $\alpha_{k+1} := \alpha_k$, increase penalty parameter $\rho_{k+1} := \theta \rho_k$.*

5. *If the stopping criterion is not satisfied set $k := k+1$ and go to step 1.*

We will define later what it needs for a step to be successful. Please note that we only decrease the regularization parameter α_k if the algorithm produces a successful step. Let us restate the system $(P_{\alpha,\rho,\mu}^k)$ that is solved by $(\bar{u}_k, \bar{y}_k, \bar{p}_k, \bar{\lambda}_k, \bar{\mu}_k)$:

$$\begin{cases} A\bar{y}_k = \bar{u}_k & \text{in } \Omega, \\ \bar{y}_k = 0 & \text{on } \partial\Omega, \end{cases} \tag{6.7a}$$

$$\begin{cases} A^*\bar{p}_k = \bar{y}_k - y_d + \bar{\mu}_k & \text{in } \Omega, \\ \bar{p}_k = 0 & \text{on } \partial\Omega, \end{cases} \tag{6.7b}$$

$$\bar{u}_k \in U_{\text{ad}}, \tag{6.7c}$$

$$(\bar{p}_k + \alpha_k \bar{u}_k + \beta\bar{\lambda}_k, u - \bar{u}_k)_{L^2(\Omega)} \geq 0 \quad \forall u \in U_{\text{ad}}, \tag{6.7d}$$

$$\bar{\mu}_k := (\mu_k + \rho_k(\bar{y}_k - \psi))_+. \tag{6.7e}$$

6.4.3 The Multiplier Update Rule

Let us start this section with a basic estimate, which will be useful in the subsequent analysis.

Lemma 6.4.4. *Let $\alpha_k > 0$ be given and let $(u^{\alpha_k}, y^{\alpha_k}, p^{\alpha_k}, \lambda^{\alpha_k}, \mu^{\alpha_k})$ be the solution of (6.4) and let $(\bar{u}_k, \bar{y}_k, \bar{p}_k, \bar{\lambda}_k, \bar{\mu}_k)$ solve (6.7). Then it holds*

$$\|y^{\alpha_k} - \bar{y}_k\|_{L^2(\Omega)}^2 + \alpha_k \|u^{\alpha_k} - \bar{u}_k\|_{L^2(\Omega)}^2 \leq (\bar{\mu}_k, \psi - \bar{y}_k)_{L^2(\Omega)} + \langle \mu^{\alpha_k}, \bar{y}_k - \psi \rangle.$$

Proof. Using (6.4c) and (6.7d), we obtain

$$\begin{aligned}
0 &\leq (p^{\alpha_k} - \bar{p}_k + \alpha_k(u^{\alpha_k} - \bar{u}_k) + \beta(\lambda^{\alpha_k} - \bar{\lambda}_k), \bar{u}_k - u^{\alpha_k})_{L^2(\Omega)} \\
&= (S^*(Su^{\alpha_k} - S\bar{u}_k), \bar{u}_k - u^{\alpha_k})_{L^2(\Omega)} - \alpha_k(\bar{u}_k - u^{\alpha_k}, \bar{u}_k - u^{\alpha_k})_{L^2(\Omega)} \\
&\quad + (S^*(\mu^{\alpha_k} - \bar{\mu}_k), \bar{u}_k - u^{\alpha_k})_{L^2(\Omega)} + \beta(\lambda^{\alpha_k} - \bar{\lambda}_k, \bar{u}_k - u^{\alpha_k})_{L^2(\Omega)}.
\end{aligned}$$

Now we use that the subdifferential is a monotone operator, which yields $(\lambda^{\alpha_k} - \bar{\lambda}_k, \bar{u}_k - u^{\alpha_k})_{L^2(\Omega)} \leq 0$. Note that $\lambda^{\alpha_k} \in \partial j(u^{\alpha_k})$ and $\bar{\lambda}_k \in \partial j(\bar{u}_k)$. This yields

$$\|y^{\alpha_k} - \bar{y}_k\|_{L^2(\Omega)}^2 + \alpha_k \|\bar{u}_k - u^{\alpha_k}\|_{L^2(\Omega)}^2 \leq (\bar{\mu}_k - \mu^{\alpha_k}, y^{\alpha_k} - \bar{y}_k). \quad (6.8)$$

The term on the right-hand side of equation (6.8) can be split into the two parts

$$(\bar{\mu}_k, y^{\alpha_k} - \bar{y}_k)_{L^2(\Omega)} = (\bar{\mu}_k, y^{\alpha_k} - \psi)_{L^2(\Omega)} + (\bar{\mu}_k, \psi - \bar{y}_k)_{L^2(\Omega)} \leq (\bar{\mu}_k, \psi - \bar{y}_k)_{L^2(\Omega)} \quad (6.9)$$

and

$$-\langle \mu_k^{\alpha_k}, y^{\alpha_k} - \bar{y}_k \rangle = -\langle \mu^{\alpha_k}, y^{\alpha_k} - \psi \rangle - \langle \mu^{\alpha_k}, \psi - \bar{y}_k \rangle = \langle \mu^{\alpha_k}, \bar{y}_k - \psi \rangle. \quad (6.10)$$

Here, we used the complementarity relation (6.4d) as well as $y^{\alpha_k} \leq \psi$ and $\bar{\mu}_k \geq 0$. Putting the inequalities (6.8), (6.9), and (6.10) together, we get

$$\|y^{\alpha_k} - \bar{y}_k\|_{L^2(\Omega)}^2 + \alpha_k \|u^{\alpha_k} - \bar{u}_k\|_{L^2(\Omega)}^2 \leq (\bar{\mu}_k, \psi - \bar{y}_k)_{L^2(\Omega)} + \langle \mu^{\alpha_k}, y_k - \psi \rangle,$$

which is the claim. $\qquad \square$

The following result motivates the update rule.

Lemma 6.4.5. *Let $(u^{\alpha_k}, y^{\alpha_k}, p^{\alpha_k}, \lambda^{\alpha_k}, \mu^{\alpha_k})$ and $(\bar{u}_k, \bar{y}_k, \bar{p}_k, \bar{\lambda}_k, \bar{\mu}_k)$ be given as in Lemma 6.4.4. Then it holds*

$$\begin{aligned}
\frac{1}{\alpha_k} \|y^{\alpha_k} - \bar{y}_k\|_{L^2(\Omega)}^2 + \|u^{\alpha_k} - \bar{u}_k\|_{L^2(\Omega)}^2 &\leq \frac{c}{\alpha_k} \Big(\|(\bar{y}_k - \psi)_+\|_{C(\bar{\Omega})} \\
&\quad + |(\bar{\mu}_k, \psi - \bar{y}_k)_{L^2(\Omega)}| \Big).
\end{aligned}$$

Proof. We start with the estimate

$$\langle \mu^{\alpha_k}, \bar{y}_k - \psi \rangle \leq \|\mu^{\alpha_k}\|_{\mathcal{M}(\Omega)} \|(\bar{y}_k - \psi)_+\|_{C(\bar{\Omega})}.$$

The result now follows using the uniform boundedness of μ^{α_k}, see Lemma 6.3.4 and Lemma 6.4.4. $\qquad \square$

This result shows that the iterates $(\bar{u}_k, \bar{y}_k)$ will converge to the solution of the regularized problem for fixed α_k if the quantity

$$\frac{1}{\alpha_k}\big(\|(\bar{y}_k - \psi)_+\|_{C(\bar{\Omega})} + |(\bar{\mu}_k, \psi - \bar{y}_k)_{L^2(\Omega)}|\big)$$

tends to zero for $k \to \infty$. To construct our update rule we follow the idea presented in [59] and define a step of Algorithm 6.5 to be successful if the condition

$$\frac{1}{\alpha_k}\big(\|(\bar{y}_k - \psi)_+\|_{C(\bar{\Omega})} + |(\bar{\mu}_k, \psi - \bar{y}_k)_{L^2(\Omega)}|\big)$$
$$\leq \frac{\tau}{\alpha_n}\big(\|(\bar{y}_n - \psi)_+\|_{C(\bar{\Omega})} + |(\bar{\mu}_n, \psi - \bar{y}_n)_{L^2(\Omega)}|\big)$$

is satisfied with $\tau \in (0,1)$. Here, we denoted by step n, $n < k$, the previous successful step. In [59] this quantity was also used as a stopping criterion. However this is not possible here, as we proceed to let α_k go to 0. Instead we will check the first order optimality conditions for problem (P) as a stopping criterion. This will be described in detail in Chapter 7.

6.4.4 The Augmented Lagrange Algorithm in Detail

Let us now formulate the algorithm based on the update rule established in the previous section.

Algorithm 6.6. *Let $\alpha_1 > 0$, $\rho_1 > 0$ and $\mu_1 \in L^2(\Omega)$ be given with $\mu_1 \geq 0$. Choose $\theta > 1, 0 < \omega < 1, \tau \in (0,1)$. Set $k := 1$ and $n := 1$.*

1. *Solve $(P^k_{\alpha,\rho,\mu})$ and obtain $(\bar{u}_k, \bar{y}_k, \bar{p}_k, \bar{\lambda}_k)$.*

2. *Set $\bar{\mu}_k := (\mu_k + \rho_k(\bar{y}_k - \psi))_+$.*

3. *Compute $R_k := \frac{1}{\alpha_k}\big(\|(\bar{y}_k - \psi)_+\|_{C(\bar{\Omega})} + |(\bar{\mu}_k, \psi - \bar{y}_k)_{L^2(\Omega)}|\big)$.*

4. *If $k = 1$ or $R_k \leq \tau R^+_{n-1}$ then the step k is successful, set*

$$\begin{cases} \alpha_{k+1} := \omega\alpha_k, \\ \mu_{k+1} := \bar{\mu}_k, \\ \rho_{k+1} := \rho_k, \end{cases}$$

and define $(u^+_n, y^+_n, p^+_n, \lambda^+_n) := (\bar{u}_k, \bar{y}_k, \bar{p}_k, \bar{\lambda}_k)$, as well as $\mu^+_n := \mu_{k+1}$, $R^+_n := R_k$ and $\alpha^+_n := \alpha_k$. Set $n := n + 1$.

5. *Otherwise if the step k is not successful, set $\mu_{k+1} := \mu_k$ and $\alpha_{k+1} := \alpha_k$, and increase the penalty parameter $\rho_{k+1} := \theta\rho_k$.*

6. *If a stopping criterion is satisfied stop, otherwise set $k := k + 1$ and go to step 1.*

Again, please note that the regularization parameter α_k is only decreased when the algorithm produces a successful step. We will take advantage of this in the subsequent analysis. Furthermore note that we set the first step to be successful. From a theoretical point this is not needed and we could just set a $R_0^+ > 0$ and test our first iteration $R_1 < \tau R_0^+$. However, if this value is chosen too small it may take several iterations until a successful step is obtained, leading to a big value of the penalization parameter ρ. As mentioned in Chapter 7 this lead to numerical instabilities.

Therefore we set the first iteration to be a successful step. This setting turned out to be effective as the new multiplier μ_1 is, in general, a better approximation than the initial guess.

6.4.5 Infinitely Many Successful Steps

The main aim of this subsection is to prove that the proposed algorithm produces infinitely many successful steps. In order to prove this we consider the augmented Lagrange KKT system of the minimization problem

$$\text{Minimize } J_\rho^\alpha(y, u, \mu) = \frac{1}{2}\|y - y_d\|_{L^2(\Omega)}^2 + \beta\|u\|_{L^1(\Omega)} + \frac{\alpha}{2}\|u\|_{L^2(\Omega)}^2 + P(y, \rho, \mu)$$

subject to $y = Su$ and $u \in U_{\text{ad}}$. We fix the multiplier approximation μ, the regularization parameter α and let the penalization parameter ρ tend to infinity. As mentioned in [59] the problem reduces to a penalty method with additional shift parameter μ. The only difference to the approach in [59] is, that we have an additional L^1-term in the objective functional. However, taking a closer look at [59, Lemma 3.6] reveals that it also holds for an additional L^1-term. The reason is that the proofs are based on an estimate similar to the one established in Lemma 6.4.4. However, there the L^1-term is already eliminated due to the monotonicity of the subdifferential. This yields the following Lemma.

Lemma 6.4.6. *Let $\mu \in L^2(\Omega)$ with $\mu \geq 0$ and $\alpha > 0$ be given. Let $(u_\rho^\alpha, y_\rho^\alpha, p_\rho^\alpha)$ be solutions of $(P_{\alpha,\rho,\mu})$ with $\rho > 0$ and (u^α, y^α) be the solution of (P^α). Then it holds $u_\rho^\alpha \to u^\alpha$ in $L^2(\Omega)$ and $y_\rho^\alpha \to y^\alpha$ in $H_0^1(\Omega) \cap C(\bar{\Omega})$ for $\rho \to \infty$.*

With a similar argument we can establish the next lemma. Again the proof can be found in [59, Lemma 3.7].

Lemma 6.4.7. *Under the same assumptions as in Lemma 6.4.6, it holds*

$$\lim_{\rho \to \infty} (\mu_\rho^\alpha, \psi - y_\rho^\alpha)_{L^2(\Omega)} = 0.$$

If we now combine these two results we can show that our algorithm produces infinitely many successful steps. This will be crucial in the convergence analysis in the next section.

Lemma 6.4.8. *Algorithm 6.6 makes infinitely many successful steps.*

Proof. We assume that the algorithm produces only finitely many successful steps. Then there is an index m such that all steps $k > m$ are not successful. Due to the definition of the algorithm we obtain $\bar{\mu}_k = \bar{\mu}_m$ for all $k > m$ and $R_k > \tau R_m > 0$ as well as $\rho_k \to \infty$ and $\alpha_k = \alpha_m$. This now yields a contradiction as with Lemma 6.4.6 and 6.4.7 we obtain

$$0 < \limsup_{k \to \infty} R_k = \lim_{k \to \infty} \frac{1}{\alpha_k} \left(\| (\bar{y}_k - \psi)_+ \|_{C(\bar{\Omega})} + |(\bar{\mu}_k, \psi - \bar{y}_k)_{L^2(\Omega)}| \right) = 0.$$

Please note that α_k is constant for $k > m$ since its value is only decreased in a successful step. □

6.5 Convergence Results

In this section we want to show convergence of Algorithm 6.6. Let us recall that the sequence $(u_n^+, y_n^+, p_n^+, \lambda_n^+)$ denotes the solution of the n-th successful iteration of Algorithm 6.6 with μ_n^+ being the corresponding approximation of the Lagrange multiplier. We start with proving L^1-boundedness of the Lagrange multipliers μ_n^+, which is accomplished in Lemma 6.5.2 below. To prove this result we need an auxiliary estimate first.

Lemma 6.5.1. *Let y_n^+, μ_n^+ be given as defined in Algorithm 6.6. Then it holds*

$$\frac{1}{\alpha_n^+} \left| (\mu_n^+, \psi - y_n^+)_{L^2(\Omega)} \right|$$

$$\leq \frac{\tau^{n-1}}{\alpha_1^+} \left(\| (y_1^+ - \psi)_+ \|_{C(\bar{\Omega})} + \| \mu_1^+ \|_{L^2(\Omega)} \| (\psi - y_1^+)_+ \|_{L^2(\Omega)} \right). \tag{6.11}$$

Proof. Using the definition for a successful step we obtain

$$\frac{1}{\alpha_n^+} |(\mu_n^+, \psi - y_n^+)_{L^2(\Omega)}|$$

$$\leq \frac{\tau}{\alpha_{n-1}^+} \left(\| (y_{n-1}^+ - \psi)_+ \|_{C(\bar{\Omega})} + |(\mu_{n-1}^+, \psi - y_{n-1}^+)_{L^2(\Omega)}| \right)$$

$$\quad - \frac{1}{\alpha_n^+} \| (y_n^+ - \psi)_+ \|_{C(\bar{\Omega})}$$

$$\leq \frac{\tau}{\alpha_{n-1}^+} \| (y_{n-1}^+ - \psi)_+ \|_{C(\bar{\Omega})} + \tau \left(\frac{1}{\alpha_{n-1}^+} |(\mu_{n-1}^+, \psi - y_{n-1}^+)_{L^2(\Omega)}| \right).$$

We now use that the $(n-1)$-th step was also successful and obtain

$$\frac{1}{\alpha_n^+}|(\mu_n^+,\psi-y_n^+)_{L^2(\Omega)}|$$

$$\leq \frac{\tau}{\alpha_{n-1}^+}\|(y_{n-1}^+-\psi)_+\|_{C(\bar\Omega)}$$

$$+\tau\left(\frac{\tau}{\alpha_{n-2}^+}\left(\|(y_{n-2}^+-\psi)_+\|_{C(\bar\Omega)}+|(\mu_{n-2}^+,\psi-y_{n-2}^+)_{L^2(\Omega)}|\right)\right.$$

$$\left.-\frac{1}{\alpha_{n-1}^+}\|(y_{n-1}^+-\psi)_+\|_{C(\bar\Omega)}\right)$$

$$\leq \frac{\tau^2}{\alpha_{n-2}^+}\left(\|(y_{n-2}^+-\psi)_+\|_{C(\bar\Omega)}+|(\mu_{n-2}^+,\psi-y_{n-2}^+)_{L^2(\Omega)}|\right).$$

The rest now follows by induction and a standard estimate. $\qquad\square$

We want to point out that the right hand side of (6.11) goes to 0 as $n\to\infty$. This will be crucial in the following convergence analysis and is a result of our update rule. Let us now show the L^1-boundedness of the Lagrange multipliers (μ_n^+).

Lemma 6.5.2. *Let Assumption 6.2.5 be fulfilled. Then Algorithm 6.6 generates an infinite sequence of bounded iterates, i.e., there is a constant $C>0$ such that for all n it holds*

$$\|y_n^+\|_{H^1(\Omega)}+\|y_n^+\|_{C(\bar\Omega)}+\|u_n^+\|_{L^2(\Omega)}+\|p_n^+\|_{W^{1,s}(\Omega)}+\|\mu_n^+\|_{L^1(\Omega)}\leq C.$$

Proof. Let $(\hat u,\hat y)$ be the Slater point given by Assumption 6.2.5, i.e., there exists $\sigma>0$ such that $\hat y+\sigma\leq\psi$. Then we can estimate

$$\sigma\|\mu_n^+\|_{L^1(\Omega)}=\int_\Omega\sigma\mu_n^+\,dx\leq\int_\Omega\mu_n^+(\psi-\hat y)\,dx=\int_\Omega\mu_n^+(\psi-y_n^++y_n^+-\hat y)\,dx$$

$$=\underbrace{\int_\Omega\mu_n^+(\psi-y_n^+)\,dx}_{(I)}+\underbrace{\int_\Omega\mu_n^+(y_n^+-\hat y)\,dx}_{(II)}.$$

The first part (I) can be estimated with Lemma 6.5.1 yielding

$$(I)\leq|(\mu_n^+,\psi-y_n^+)_{L^2(\Omega)}|\leq\frac{\alpha_n^+}{\alpha_1^+}\tau^{n-1}\left(\|(y_1^+-\psi)_+\|_{C(\bar\Omega)}\right.$$

$$\left.+\|\mu_1^+\|_{L^2(\Omega)}\|(\psi-y_1^+)_+\|_{L^2(\Omega)}\right)$$

$$\leq c\tau^{n-1}.$$

$$(6.12)$$

Please note that we used the monotonicity of $(\alpha_n)_n$. Before we estimate part (II), recall that we have the inequality

$$(\lambda_n^+,u-u_n^+)_{L^2(\Omega)}\leq\|u\|_{L^1(\Omega)}-\|u_n^+\|_{L^1(\Omega)}$$

for every $u \in L^1(\Omega)$. By definition we obtain that $u \in U_{ad}$ implies $u \in L^\infty(\Omega)$. Now the second part (II) can be estimated using Young's Inequality as follows

$$
\int_\Omega \mu_n^+ (y_n^+ - \hat{y}) \, dx = \langle A^* p_n^+ - (y_n^+ - y_d), y_n^+ - \hat{y} \rangle
$$

$$
= \langle p_n^+, A(y_n^+ - \hat{y}) \rangle - (y_n^+ - y_d, y_n^+ - \hat{y})_{L^2(\Omega)}
$$

$$
= (p_n^+, u_n^+ - \hat{u})_{L^2(\Omega)} - (y_n^+ - y_d, y_n^+ - \hat{y})_{L^2(\Omega)}
$$

$$
\leq -(\alpha_n^+ u_n^+, u_n^+ - \hat{u})_{L^2(\Omega)} - (y_n^+ - y_d, y_n^+ - \hat{y})_{L^2(\Omega)} - \beta(\lambda_n^+, u_n^+ - \hat{u})_{L^2(\Omega)}
$$

$$
\leq c.
$$

$$
(6.13)
$$

By our assumption we know that u_n^+ and y_n^+ are uniformly bounded and the subgradient $\lambda_n^+ \in L^\infty(\Omega)$ is uniformly bounded by construction.
Combining (6.12) and (6.13) yields

$$
\|\mu_n^+\|_{L^1(\Omega)} \leq c \frac{\tau^{n-1}}{\sigma} + \frac{c}{\sigma}.
$$

Since $\tau \in (0,1)$ by assumption, the right-hand side is bounded. The result now follows from Theorem 6.4.3. $\qquad\square$

Theorem 6.5.3. *As $n \to \infty$ we have for the sequence (u_n^+, y_n^+) generated by Algorithm 6.6*

$$
(u_n^+, y_n^+) \to (\bar{u}, \bar{y}), \ in \ L^2(\Omega) \times \left(H_0^1(\Omega) \cap C(\bar{\Omega}) \right),
$$

where $\bar{u}$ denotes the unique solution of (P) with associated state $\bar{y}$.

Proof. Since the algorithm yields an infinite number of successful steps, see Lemma 6.4.8 we get

$$
\lim_{n\to\infty} R_n^+ = \lim_{n\to\infty} \frac{1}{\alpha_n^+} \left(\|(y_n^+ - \psi)_+\|_{C(\bar{\Omega})} + |(\mu_n^+, \psi - y_n^+)_{L^2(\Omega)}| \right) = 0, \quad (6.14)
$$

with $\alpha_n^+ \to 0$. Let $(u^{\alpha_n^+}, y^{\alpha_n^+}, p^{\alpha_n^+}, \lambda^{\alpha_n^+}, \mu^{\alpha_n^+})$ be a solution of (6.4) for $\alpha := \alpha_n^+$ then we obtain from Lemma 6.4.4 the following inequality

$$
\frac{1}{\alpha_n^+} \left\| y^{\alpha_n^+} - y_n^+ \right\|_{L^2(\Omega)}^2 + \left\| u^{\alpha_n^+} - u_n^+ \right\|_{L^2(\Omega)}^2
$$

$$
\leq \frac{1}{\alpha_n^+} \left(\langle \mu^{\alpha_n^+}, y_n^+ - \psi \rangle + |(\mu_n^+, \psi - y_n^+)_{L^2(\Omega)}| \right)
$$

$$
\leq \frac{1}{\alpha_n^+} \left(\left\| \mu^{\alpha_n^+} \right\|_{\mathcal{M}(\bar{\Omega})} \|(y_n^+ - \psi)_+\|_{C(\bar{\Omega})} + |(\mu_n^+, \psi - y_n^+)_{L^2(\Omega)}| \right)
$$

$$
\leq \frac{c}{\alpha_n^+} \left(\|(y_n^+ - \psi)_+\|_{C(\bar{\Omega})} + |(\mu_n^+, \psi - y_n^+)_{L^2(\Omega)}| \right).
$$

Note that in the last step we used Lemma 6.3.4. With (6.14) from above, we conclude

$$\lim_{n \to \infty} \frac{1}{\alpha_n^+} \left\| y^{\alpha_n^+} - y_n^+ \right\|_{L^2(\Omega)}^2 + \left\| u^{\alpha_n^+} - u_n^+ \right\|_{L^2(\Omega)}^2 = 0. \tag{6.15}$$

As $\alpha_n^+ \to 0$ with $n \to \infty$ we obtain by Theorem 6.3.1 that $u^{\alpha_n^+} \to \bar{u}$. Triangular inequality now reveals

$$\|u_n^+ - \bar{u}\|_{L^2(\Omega)} \le \|u_n^+ - u^{\alpha_n^+}\|_{L^2(\Omega)} + \|u^{\alpha_n^+} - \bar{u}\|_{L^2(\Omega)} \to 0.$$

Convergence of $y_n^+ \to \bar{y}$ follows from Theorem 6.2.2 which finishes the proof. $\qquad \square$

Remark 6.5.4. *For the sequence (y_n^+) generated by Algorithm 6.6 we obtain*

$$\frac{1}{\alpha_n^+} \|y_n^+ - \bar{y}\|_{L^2(\Omega)}^2 \to 0,$$

which is similar to the results obtained for a Tikhonov regularization without state constraints, see [97].

Proof. We split the error to obtain with some $c > 0$ independent from n

$$\frac{1}{\alpha_n^+} \|y_n^+ - \bar{y}\|_{L^2(\Omega)}^2 \le \frac{c}{\alpha_n^+} \|y_n^+ - y^{\alpha_n^+}\|_{L^2(\Omega)}^2 + \frac{c}{\alpha_n^+} \|y^{\alpha_n^+} - \bar{y}\|_{L^2(\Omega)}^2.$$

The results is now an immediate consequence of (6.15) and Theorem 6.3.1. $\qquad \square$

Since Lagrange multipliers are in general not uniquely given, we cannot expect weak or even strong convergence of $(p_n^+)_n$ or $(\mu_n^+)_n$ in their respective spaces. Let us therefore now assume that the adjoint state $\bar{p}$ and the multiplier corresponding to the state constraint $\bar{\mu}$ are unique. Then following [59, Theorem 3.12, Corollary 3.12] we get the following two results.

Lemma 6.5.5. *Let $s \in (1, \frac{N}{N-1})$ such that the embedding $W^{1,s}(\Omega) \to L^2(\Omega)$ is compact. Let subsequences $(u_{n_j}^+, y_{n_j}^+, p_{n_j}^+, \lambda_{n_j}^+, \mu_{n,j}^+)_j$ of $(u_n^+, y_n^+, p_n^+, \lambda_n^+, \mu_n^+)_n$ be given such that $\mu_{n_j} \rightharpoonup^* \bar{\mu}$ in $\mathcal{M}(\bar{\Omega})$ and $p_{n_j}^+ \rightharpoonup \bar{p}$ in $W^{1,s}(\Omega)$. Then $(\bar{u}, \bar{y}, \bar{p}, \bar{\lambda}, \bar{\mu})$ with $\bar{\lambda}(x) := P_{[-1,1]}(-\beta^{-1}\bar{p}(x))$ satisfy the optimality system (6.3a)-(6.3d) of (P).*

Proof. Most of the proof can be found in [59, Theorem 3.12]. We only have to take care about the additional subgradient $\lambda_{n_j}^+$ and show that the variational inequality (6.3c) is satisfied. By Lemma 6.3.3 we obtain

$$\lambda_{n_j}^+(x) = P_{[-1,1]} \left(-\frac{1}{\beta}(p_{n_j}^+(x) + \alpha_{n_j}^+ u_{n_j}^+(x)) \right).$$

We now use the compact embedding $W^{1,s}(\Omega) \to L^2(\Omega)$ to obtain $p_{n_j}^+ \to \bar{p}$ in $L^2(\Omega)$ and hence $\lambda_{n_j}^+ \to \bar{\lambda}$ in $L^2(\Omega)$. Note that we also used the boundedness of $u_{n_j}^+$ and $\alpha_{n_j} \to 0$. Now we use the strong convergence of the control $u_{n_j}^+ \to \bar{u}$ in $L^2(\Omega)$ to obtain

$$(p_{n_j}^+ + \alpha_{n_j}^+ u_{n_j}^+ + \beta\lambda_{n_j}^+, u - u_{n_j}^+)_{L^2(\Omega)} \to (\bar{p} + \beta\bar{\lambda}, u - \bar{u}) \ge 0,$$

which finishes the proof. $\qquad \square$

A direct consequence is the following result.

Theorem 6.5.6. *Let $(\bar{u}, \bar{y}, \bar{p}, \bar{\lambda}, \bar{\mu})$ satisfy the KKT-system (6.3). Let us assume that $(\bar{p}, \bar{\mu})$ are uniquely given. Pick $s \in \left(1, \frac{N}{N-1}\right)$ such that the embedding $W^{1,s}(\Omega) \to L^2(\Omega)$ is compact. Then $\bar{\lambda}$ is also unique and it holds*

$$
\begin{aligned}
p_n^+ &\rightharpoonup \bar{p} && \text{in } W^{1,s}(\Omega), \\
\mu_n^+ &\overset{*}{\rightharpoonup} \bar{\mu} && \text{in } \mathcal{M}(\bar{\Omega}), \\
\lambda_n^+ &\to \bar{\lambda} && \text{in } L^2(\Omega).
\end{aligned}
$$

Let us note that finding sufficient conditions such that the multiplier are unique is a non-trivial task. For the optimal control of a semilinear partial differential equation sufficient conditions are established in [62]. There the active sets with respect to the state constraints and the control constraints have to be separated.

CHAPTER 7

Numerical Implementation of the ALM

In this chapter we want to introduce an active-set method for the solution of the subproblems arising in the augmented Lagrange method stated in Algorithm 6.6. The subproblem is given as

$$\min_{u \in U_{\mathrm{ad}}} \frac{1}{2}\|y - y_d\|^2_{L^2(\Omega)} + \beta\|u\|_{L^1(\Omega)} + \frac{\alpha}{2}\|u\|^2_{L^2(\Omega)} + P(y, \rho, \mu). \qquad (P_{\alpha,\rho,\mu})$$

The necessary optimality conditions obtained in Theorem 6.4.2 are not suited to derive a numerical method as they only introduce a multiplier for the state constraints. We follow [89] and also introduce multipliers for the bilateral inequality constraints for the control. Hence the optimal solution $(\bar{u}, \bar{y}, \bar{p}) \in L^2(\Omega) \times H_0^1(\Omega) \times H_0^1(\Omega)$ of $(P_{\alpha,\rho,\mu})$ is characterized by the existence of $\lambda, \lambda^a, \lambda^b \in L^2(\Omega)$ such that

$$\begin{cases} A\bar{y} = \bar{u} & \text{in } \Omega, \\ \bar{y} = 0 & \text{on } \partial\Omega, \end{cases} \qquad (7.1\text{a})$$

$$\begin{cases} A^*\bar{p} = \bar{y} - y_d + \bar{\mu} & \text{in } \Omega, \\ \bar{p} = 0 & \text{on } \partial\Omega, \end{cases} \qquad (7.1\text{b})$$

$$\bar{p} + \alpha\bar{u} + \lambda + \lambda^b - \lambda^a = 0, \qquad (7.1\text{c})$$

$$\lambda^a \geq 0, \quad \bar{u} - u_a \geq 0, \quad \lambda^a(\bar{u} - u_a) = 0, \qquad (7.1\text{d})$$

$$\lambda^b \geq 0, \quad u_b - \bar{u} \geq 0, \quad \lambda^b(u_b - \bar{u}) = 0, \qquad (7.1\text{e})$$

$$\begin{cases} \lambda = \beta & \text{on } \{x \in \Omega : \bar{u} > 0\}, \\ |\lambda| \leq \beta & \text{on } \{x \in \Omega : \bar{u} = 0\}, \\ \lambda = -\beta & \text{on } \{x \in \Omega : \bar{u} < 0\}, \end{cases} \qquad (7.1\text{f})$$

$$\bar{\mu} := (\mu + \rho(\bar{y} - \psi))_+ . \qquad (7.1\text{g})$$

Here (7.1a) is the state equation, (7.1b) characterizes the adjoint state, (7.1d)-(7.1e) define the multipliers for the control constraints and (7.1f) reflects the fact that $\lambda \in \partial\beta\|\bar{u}\|_{L^1(\Omega)}$.

In the following we want to formulate our active-set method and present several numerical examples for Algorithm 6.6. The implementation was done with FEniCS [64] using the DOLFIN [65] Python interface.

7.1   Active-Set Method

The arising subproblems $(P_{\alpha,\rho,\mu})$ are solved by combining two methods. The first method is the active-set method presented by Stadler [89], where optimal control problems of type $(P_{\alpha,\rho,\mu})$ were solved, but without augmented state constraints. The second is the method established by Ito and Kunisch [51] that presented an active-set method for optimal control problems with state constraints but without an L^1-cost term.

In the following u_k, y_k, p_k and λ_k are iterates generated by the active-set method, which is described below.

Like in [89] we set

$$\xi_k := \lambda_k - \lambda_k^a + \lambda_k^b,$$

where λ_k denotes the subdifferential of $\beta \|\bar{u}_k\|_{L^1(\Omega)}$, λ_k^a the multiplier to the lower control constraints $u_a - u_k \leq 0$ and λ_k^b the multiplier corresponding to the upper control constraint $u_k - u_b \leq 0$. Then (7.1c) can be written as

$$p_k + \alpha u_k + \xi_k = 0.$$

Note that if $(u_k, y_k, p_k, \lambda_k)$ are a solution of $(P_{\alpha,\rho,\mu})$ we can reconstruct the multipliers via the formula

$$
\begin{aligned}
\lambda_k &= \min(\beta, \max(-\beta, \xi_k)), \\
\lambda_k^a &= -\min(0, \xi_k + \beta), \\
\lambda_k^b &= \max(0, \xi_k - \beta).
\end{aligned}
\tag{7.2}
$$

Now define the following sets, see also Lemma 6.3.3

$$
\begin{aligned}
\mathcal{Y}_+^k &= \{x \in \Omega : (\mu + \rho(y_k - \psi)) > 0\}, \\
\mathcal{Y}_-^k &= \Omega \setminus \mathcal{Y}_+^k, \\
\mathcal{A}_a^k &= \{x \in \Omega : p_k \geq \beta - \alpha u_a\}, \\
\mathcal{A}_0^k &= \{x \in \Omega : |p_k| < \beta\}, \\
\mathcal{A}_b^k &= \{x \in \Omega : p_k \leq -\alpha u_b - \beta\}, \\
\mathcal{I}_-^k &= \{x \in \Omega : \beta \leq p_k < \beta - \alpha u_a\}, \\
\mathcal{I}_+^k &= \{x \in \Omega : -\alpha u_b - \beta < p_k \leq -\beta\}.
\end{aligned}
$$

The sets $\mathcal{A}_a^k$, $\mathcal{A}_0^k$ and $\mathcal{A}_b^k$ are called active sets, as on $\mathcal{A}_a^k$ we obtain $u_k = u_a$, on $\mathcal{A}_b^k$ we get $u_k = u_b$ and on $\mathcal{A}_0^k$ we have $u_k = 0$. Obviously, the five sets $\mathcal{A}_a^k$, $\mathcal{A}_0^k$, $\mathcal{A}_b^k$, $\mathcal{I}_-^k$ and $\mathcal{I}_+^k$ are disjoint and their union is Ω. The sets $\mathcal{Y}_-^k$ and $\mathcal{Y}_+^k$ are motivated by (7.1g).

The resulting subproblem of the augmented Lagrange method $(P_{\alpha,\rho,\mu})$ can now be solved by the following algorithm, which is an active-set method.

Algorithm 7.7. *Choose initial data u_0, p_0 and parameters α, ρ, compute the sets $\mathcal{Y}^0_-$,*
$\mathcal{Y}^0_+, \mathcal{A}^0_a, \mathcal{A}^0_0, \mathcal{A}^0_b, \mathcal{I}^0_-, \mathcal{I}^0_+$.

1. *Solve for $(u_{k+1}, y_{k+1}, p_{k+1}, \xi_{k+1})$ satisfying*

$$
\begin{aligned}
Ay_{k+1} - u_{k+1} &= 0, \\
-A^* p_{k+1} + y_{k+1} - y_d + \mu_{k+1} &= 0, \\
p_{k+1} + \alpha u_{k+1} + \xi_{k+1} &= 0,
\end{aligned}
\tag{7.3a}
$$

$$
(1 - \chi_{\mathcal{A}^k_a} - \chi_{\mathcal{A}^k_b} - \chi_{\mathcal{A}^k_0}) \xi_{k+1} + (1 - \chi_{\mathcal{I}^k_-} - \chi_{\mathcal{I}^k_+}) u_{k+1}
$$
$$
= \chi_{\mathcal{A}^k_a} u_a + \chi_{\mathcal{A}^k_b} u_b - \chi_{\mathcal{I}^k_-} \beta + \chi_{\mathcal{I}^k_+} \beta,
\tag{7.3b}
$$

$$
\mu_{k+1} = \begin{cases} 0 & \text{on } \mathcal{Y}^k_-, \\ \mu + \rho(y_{k+1} - \psi) & \text{on } \mathcal{Y}^k_+. \end{cases}
\tag{7.3c}
$$

2. *Compute the sets $\mathcal{Y}^{k+1}_-, \mathcal{Y}^{k+1}_+, \mathcal{A}^{k+1}_a, \mathcal{A}^{k+1}_0, \mathcal{A}^{k+1}_b, \mathcal{I}^{k+1}_-, \mathcal{I}^{k+1}_+$.*

3. *If the following equalities hold: $\mathcal{A}^{k+1}_a = \mathcal{A}^k_a$, $\mathcal{A}^{k+1}_0 = \mathcal{A}^k_0$, $\mathcal{A}^{k+1}_b = \mathcal{A}^k_b$, $\mathcal{I}^{k+1}_- = \mathcal{I}^k_-$, $\mathcal{I}^{k+1}_+ = \mathcal{I}^k_+$, $\mathcal{Y}^{k+1}_- = \mathcal{Y}^k_-$ and $\mathcal{Y}^{k+1}_+ = \mathcal{Y}^k_+$ then go step 4. Otherwise set $k = k + 1$ and go to step 2.*

4. *Compute the subdifferential $\lambda_{k+1} := \min(\beta, \max(-\beta, \xi_{k+1}))$ and stop the algorithm.*

Note that (7.3b) can be equivalently written as

$$
u_{k+1} = \begin{cases} u_a & \text{on } \mathcal{A}^k_a, \\ 0 & \text{on } \mathcal{A}^k_0, \\ u_b & \text{on } \mathcal{A}^k_b, \end{cases}
$$

$$
\xi_{k+1} = \begin{cases} -\beta & \text{on } \mathcal{I}^k_-, \\ \beta & \text{on } \mathcal{I}^k_+, \end{cases}
$$

but it is more accessible in this form.

The computation of the L^1-subgradient follows from a reconstruction formula, [89, Chapter 2]. Further, the termination criterion yields a solution of the augmented Lagrange subproblem $(P_{\alpha,\rho,\mu})$.

Lemma 7.1.1. *If the following equalities hold*

$$
\begin{array}{ccccccc}
A_a^{k+1} & = & A_a^k, & A_0^{k+1} & = & A_0^k, & A_b^{k+1} & = & A_b^k, & \mathcal{I}_-^{k+1} & = & \mathcal{I}_-^k, \\
\mathcal{I}_+^{k+1} & = & \mathcal{I}_+^k, & \mathcal{Y}_-^{k+1} & = & \mathcal{Y}_-^k, & \mathcal{Y}_+^{k+1} & = & \mathcal{Y}_+^k,
\end{array}
$$

then $(u_{k+1}, y_{k+1}, p_{k+1}, \mu_{k+1}, \lambda_{k+1})$ *is a solution to (6.6) with* α, μ *and* β *fixed.*

Proof. Since for given sets the solution to (7.3) is unique we have

$$
(u_{k+1}, y_{k+1}, p_{k+1}) = (u_k, y_k, p_k).
$$

By definition of the sets $\mathcal{Y}_-^{k+1}$ and $\mathcal{Y}_+^{k+1}$ we get $\mu_{k+1} = (\mu + \rho(y_{k+1} - \psi))_+$. Furthermore we know, that $(u_{k+1}, y_{k+1}, p_{k+1}, \lambda_{k+1})$ satisfy (7.1d)-(7.1f) if and only if it satisfies the nonsmooth equation, see [89, Lemma 2.2]

$$
u_{k+1} - \max\left(0, u_{k+1} + c(\xi_{k+1} - \beta)\right) - \min\left(0, u_{k+1} + c(\xi_{k+1} + \beta)\right)
$$
$$
+ \max\left(0, (u_{k+1} - u_b) + c(\xi_{k+1} - \beta)\right) + \min\left(0, (u_{k+1} - u_a) + c(\xi_{k+1} + \beta)\right) = 0,
$$

where $c > 0$ arbitrary. Choosing $c = \alpha^{-1}$ and exploiting $\xi_{k+1} = -p_{k+1} - \alpha u_{k+1}$ we get the equivalent formulation

$$
u_{k+1} - \alpha^{-1} \max(0, -p_{k+1} - \beta) - \alpha^{-1} \min(0, -p_{k+1} + \beta)
$$
$$
+ \alpha^{-1} \max(0, -p_{k+1} - \beta - \alpha u_b) + \alpha^{-1} \min(0, -p_{k+1} + \beta - \alpha u_a) = 0. \tag{7.4}
$$

Let us show that (7.4) holds on the set A_a^{k+1}. A straightforward calculation yields

$$
\alpha^{-1} \max(0, -p_{k+1} - \beta) = 0,
$$
$$
\alpha^{-1} \min(0, -p_{k+1} + \beta) = \alpha^{-1}(-p_{k+1} + \beta),
$$
$$
\alpha^{-1} \max(0, -p_{k+1} - \beta - \alpha u_b) = 0,
$$
$$
\alpha^{-1} \min(0, -p_{k+1} + \beta - \alpha u_a) = \alpha^{-1}(-p_{k+1} + \beta - \alpha u_a).
$$

Hence (7.4) holds on the A_a^{k+1}. With similar arguments it can be easily shown that (7.4) holds on the sets A_b^{k+1} and A_0^{k+1}. Let us now show it for the set $\mathcal{I}_-^k$. On this set we know by (7.3b) that $\xi_{k+1} = -\beta$ and the adjoint state satisfies $\beta \leq p_{k+1} < \beta - \alpha u_a$. Furthermore we know from (7.3a) that

$$
u_{k+1} = \frac{1}{\alpha}(-p_{k+1} - \zeta_{k+1}) = \frac{1}{\alpha}(-p_{k+1} + \beta).
$$

A short calculation reveals

$$
\alpha^{-1} \max(0, -p_{k+1} - \beta) = 0,
$$
$$
\alpha^{-1} \min(0, -p_{k+1} + \beta) = \alpha^{-1}(-p_{k+1} + \beta),
$$
$$
\alpha^{-1} \max(0, -p_{k+1} - \beta - \alpha u_b) = 0,
$$
$$
\alpha^{-1} \min(0, -p_{k+1} + \beta - \alpha u_a) = 0,
$$

which shows that (7.4) holds on $\mathcal{I}_-^k$. Again with a similar argument one can show that (7.4) also holds on $\mathcal{I}_+^k$. Recall that

$$\Omega = \mathcal{A}_a^{k+1} \cup \mathcal{A}_b^{k+1} \cup \mathcal{A}_0^{k+1} \cup \mathcal{I}_-^k \cup \mathcal{I}_+^k.$$

Consequently the functions $(u_{k+1}, y_{k+1}, p_{k+1}, \mu_{k+1}, \lambda_{k+1})$ satisfy (7.4), where the multipliers are reconstructed using (7.2). $\qquad\square$

However, high values of the penalty parameter ρ paired with small values of the Tikhonov parameter α may evoke bad stability during solution of the subproblem. To counteract this aspect we introduce a so called intermediate step. Here, Step 3 and Step 4 of Algorithm 6.6 are extended for a third alternative. If the current iterates of the k-th iteration do not satisfy the update rule but sufficiently satisfy the feasibility and complementarity condition, i.e.

$$R_k \geq \tau R_{n-1}^+ \quad \text{and} \quad \left\|(\bar{y}_k - \psi)_+\right\|_{C(\bar{\Omega})} + \left|(\bar{\mu}_k, \psi - \bar{y}_k)_{L^2(\Omega)}\right| < \varepsilon_I,$$

with $\varepsilon_I > 0$, we define this step to be almost successful and set

$$\alpha_{k+1} := \omega \alpha_k,$$
$$\mu_{k+1} := \bar{\mu}_k,$$
$$(u_n^+, y_n^+, p_n^+, \lambda_n^+, \mu_k^+, \alpha_k^+) := (\bar{u}_k, \bar{y}_k, \bar{p}_k, \bar{\lambda}_k, \bar{\mu}_k, \alpha_k).$$

Such a step is called an intermediate step.

As a termination criterion we check the optimality conditions of the current iterate $(u_n^+, y_n^+, p_n^+, \mu_n^+, \lambda_n^+)$ i.e. we stop the algorithm if the inequality

$$\left\|u_n^+ - P_{U_{ad}}\left(u_n^+ - (p_n^+ + \beta\lambda_n^+)\right)\right\|_{L^2(\Omega)}$$
$$+ \left\|(y_n^+ - \psi)_+\right\|_{C(\bar{\Omega})} + \left|(\mu_n^+, y_n^+ - \psi)_{L^2(\Omega)}\right| \leq \varepsilon$$

is satisfied. In order to be consistent we set $\varepsilon_I < \varepsilon$.

As the active-set methods are related to the class of semi-smooth Newton methods we cannot expect a global convergence behavior of the method described above. Furthermore, the problem becomes badly conditioned if $\alpha \to 0$ or $\rho \to \infty$. Due to the intermediate step we expect ρ to be bounded.

7.2 Numerical Results

Let us present some numerical results to support our method. We apply our method for problems of the following form:

$$\min\ J(y, u) := \frac{1}{2}\|y - y_d\|_{L^2(\Omega)}^2 + \beta\|u\|_{L^1(\Omega)}$$

subject to

$$
\begin{aligned}
Ay &= u + f && \text{in } \Omega, \\
y &= 0 && \text{on } \partial\Omega, \\
y &\le \psi && \text{in } \Omega, \\
u_a &\le u \le u_b && \text{in } \Omega.
\end{aligned}
$$

The additional variable $f \in L^2(\Omega)$ allows us to construct test problems with known solutions. Note that this problem is of form (P).

7.2.1 Bang-Bang-Off Solution in One Space Dimension

We first consider the one-dimensional case and define $\Omega = (-1, 1)$, $u_a = -1$, $u_b = 1$ and $\beta = 1$. Furthermore set

$$
\bar{y}(x) := \begin{cases}
28 + 108 \cdot x + 144 \cdot x^2 + 64 \cdot x^3 & \text{if } x \in [-1, -\frac{3}{4}], \\
1 & \text{if } x \in [-\frac{3}{4}, \frac{3}{4}], \\
28 - 108 \cdot x + 144 \cdot x^2 - 64 \cdot x^3 & \text{if } x \in [\frac{3}{4}, 1],
\end{cases}
$$

$$
\bar{p}(x) := -2\cos\left(\frac{3\pi}{2}x\right),
$$

$$
\bar{u}(x) := \begin{cases}
0 & \text{if } x \in [-1, -\frac{8}{9}] \cup [-\frac{4}{9}, -\frac{2}{9}] \cup [\frac{2}{9}, \frac{4}{9}] \cup [\frac{8}{9}, 1], \\
1 & \text{if } x \in (-\frac{2}{9}, \frac{2}{9}), \\
-1 & \text{if } x \in (-\frac{8}{9}, -\frac{4}{9}) \cup (\frac{4}{9}, \frac{8}{9}),
\end{cases}
$$

$$
\bar{\mu}(x) := \begin{cases}
\mathrm{Exp}\left(-\frac{1}{1-(\frac{4}{3}x)^2}\right) & \text{if } x \in [-\frac{3}{4}, \frac{3}{4}], \\
0 & \text{else},
\end{cases}
$$

$$
\psi(x) := 1.
$$

Some calculations show that $\bar{y}, \bar{p} \in C^2(\Omega)$ and $\bar{y} = \bar{p} = 0$ on $\partial\Omega$. By construction we obtain $\bar{u}(x) \in \{-1, 0, 1\}$ for a.e. $x \in \Omega$. In order to satisfy the optimality conditions we now set

$$
\begin{aligned}
f(x) &:= -\Delta\bar{y}(x) - \bar{u}(x), \\
y_d(x) &:= \Delta\bar{p}(x) + \bar{y}(x) + \bar{\mu}(x).
\end{aligned}
$$

The functions $(\bar{u}, \bar{y}, \bar{p}, \bar{\mu})$ satisfy the KKT conditions defined in Theorem 6.3.2 with a suitable modification for the forward equation. We apply our algorithm with the following set of parameters

$$
\theta = 5, \quad \omega = 0.75, \quad \tau = 0.8, \quad \varepsilon = 10^{-6}, \quad \varepsilon_I = 5 \cdot 10^{-7}.
$$

The interval Ω is divided into 10^6 equidistant elements. The algorithm stops after a total of 40 iterations, which splits in 13 successful, 19 intermediate and 8 unsuccessful iterations with an average of 5.25 inner iterations. The parameters were initialized with $\alpha := 1$ and $\rho := 100$ and the final parameters are $\alpha = 0.75^{32} \approx 10^{-4}$ and $\rho = 100 \cdot 5^8 \approx 3.9 \cdot 10^7$.

As we have an exact solution we can compute convergence rates. We plot the L^2-error $\|u_k^+ - \bar{u}\|_{L^2(\Omega)}$ over the regularization parameter α_k. Note that we only plot successful and intermediate steps. As expected we see that the algorithm produces only intermediate steps after some given time. The error can be found in Figure 7.1 and plots of the computed solution can be seen in Figure 7.2.

Remark 7.2.1. *Analysing the error $\|u_k^+ - \bar{u}\|_{L^2(\Omega)}$ we see that the error behaves like*

$$\|u_k^+ - \bar{u}\|_{L^2(\Omega)} = \mathcal{O}\left((\alpha_k^+)^{\frac{1}{2}}\right). \tag{7.5}$$

We want to mention that the exact control $\bar{u}$ satisfies the following regularity assumption $\operatorname{meas}\{x \in \Omega : \left|\,|\bar{p}(x)| - \beta\,\right| < \varepsilon\} \leq c\varepsilon$, *for all $\varepsilon > 0$ with some $\kappa > 0$, which resembles Assumption (ASC). In fact we will use this regularity assumption in Chapter 9 to prove regularization error estimates for a Tikhonov regularization.*
Furthermore we used the similar Assumption (ASC) to prove convergence rates for the iterative Bregman regularization method, see Chapter 3. However, it is an open problem to prove convergence rates for Algorithm 6.6 presented in Chapter 6.

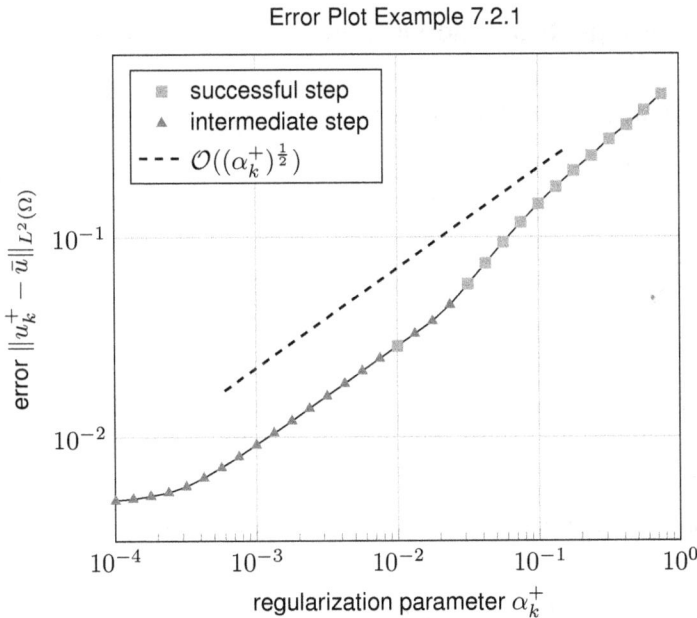

Figure 7.1: Error $\|u_k^+ - \bar{u}\|_{L^2(\Omega)}$ over the regularization parameter α_k^+ for Example 7.2.1.

(a) Computed control u

(b) Computed state y

(c) Computed multiplier μ

(d) Computed adjoint state p

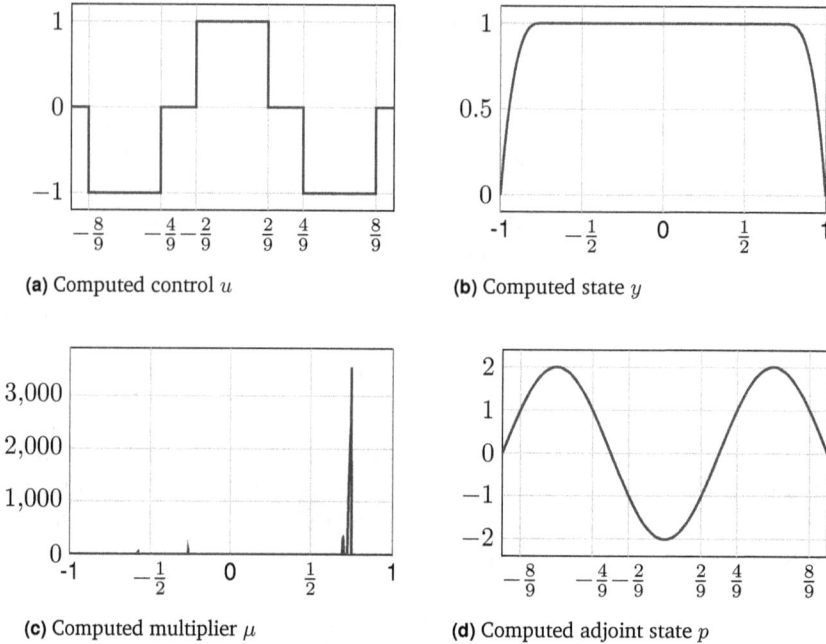

Figure 7.2: Computed results for Example 7.2.1.

7.2.2 Bang-Bang-Off Solution in Two Space Dimensions

We set $u_a = -1$, $u_b = 1$. Let Ω be the circle around 0 with radius 2. We now define the following functions. To shorten our notation we set $r := r(x, y) := \sqrt{x^2 + y^2}$.

$$\bar{y}(x, y) := \begin{cases} 1 & \text{if } r < 1, \\ 32 - 120 \cdot r + 180 \cdot r^2 - 130 \cdot r^3 + 45 \cdot r^4 - 6 \cdot r^5 & \text{if } r \geq 1, \end{cases}$$

$$\bar{p}(x, y) := \sin(x) \cdot \sin(y) \cdot \left(1 - \frac{5}{4}r^3 + \frac{15}{16}r^4 - \frac{3}{16}r^5\right),$$

$$\bar{u}(x, y) := -\mathrm{Sign}(\bar{p}(x, y)),$$

$$\bar{\mu}(x, y) := \begin{cases} \mathrm{Exp}\left(-\frac{1}{1-r^2}\right) & \text{if } r < 1, \\ 0 & \text{if } r \geq 1, \end{cases}$$

$$\psi(x, y) := 1.$$

Some calculation show that $\bar{\mu}, \bar{p} \in C^2(\bar{\Omega})$ and $\bar{\mu} \in C(\bar{\Omega})$. Furthermore $\bar{y} = \bar{p} = 0$ on $\partial\Omega$. We now set

$$f(x, y) := -\Delta\bar{y}(x, y) - \bar{u}(x, y),$$
$$y_d(x, y) := \Delta\bar{p}(x, y) + \bar{y}(x, y) + \bar{\mu}(x, y).$$

(a) Computed control for $\beta = 0.05$

(b) Computed control for $\beta = 0.1$

(c) Computed control for $\beta = 0.2$

(d) Computed control for $\beta = 1$

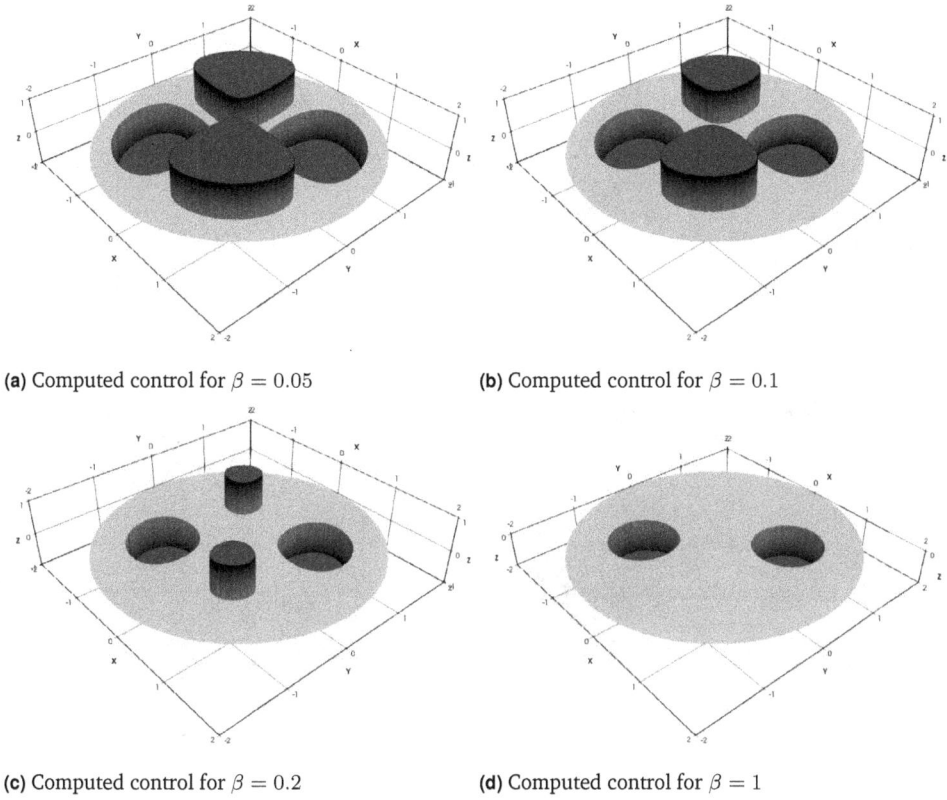

Figure 7.3: Computed discrete control for Example 7.2.2 for different values of β.

One now can check that for $\beta = 0$ the functions $(\bar{u}, \bar{y}, \bar{p}, \bar{\mu})$ satisfy the KKT conditions defined in Theorem 6.3.2 leading to a bang-bang solution. For $\beta \neq 0$ we expect the optimal solution to exhibit a bang-bang-off structure. Here no exact solution is known. We computed this problem for different values of β on a regular triangular grid with approximately $1.8 \cdot 10^5$ degrees of freedom. The parameters used for this computation are $\tau = 0.8$, $\omega = 0.75$, $\theta = 10$, $\varepsilon = 10^{-6}$ and $\varepsilon_I = 5 \cdot 10^{-7}$. We started with $\alpha = 0.1$ and $\rho = 100$. Additional information for the calculations can be found in Table 7.1 while the computed controls can be seen in Figure 7.3.

As expected by Theorem 6.2.8 we observe that the solution becomes more sparse as β becomes large. Taking a look at the final values of the regularization parameter α and penalization parameter ρ we see, that they are of the same order of magnitude for all β. However for bigger β more successful steps are obtained. The number of unsuccessful steps is nearly constant. This supports the introduction of the intermediate step, as this controls the penalization parameter and stabilizes the algorithm.

(a) Computed control u

(b) Computed state y

(c) Computed adjoint state p

(d) Computed multiplier μ

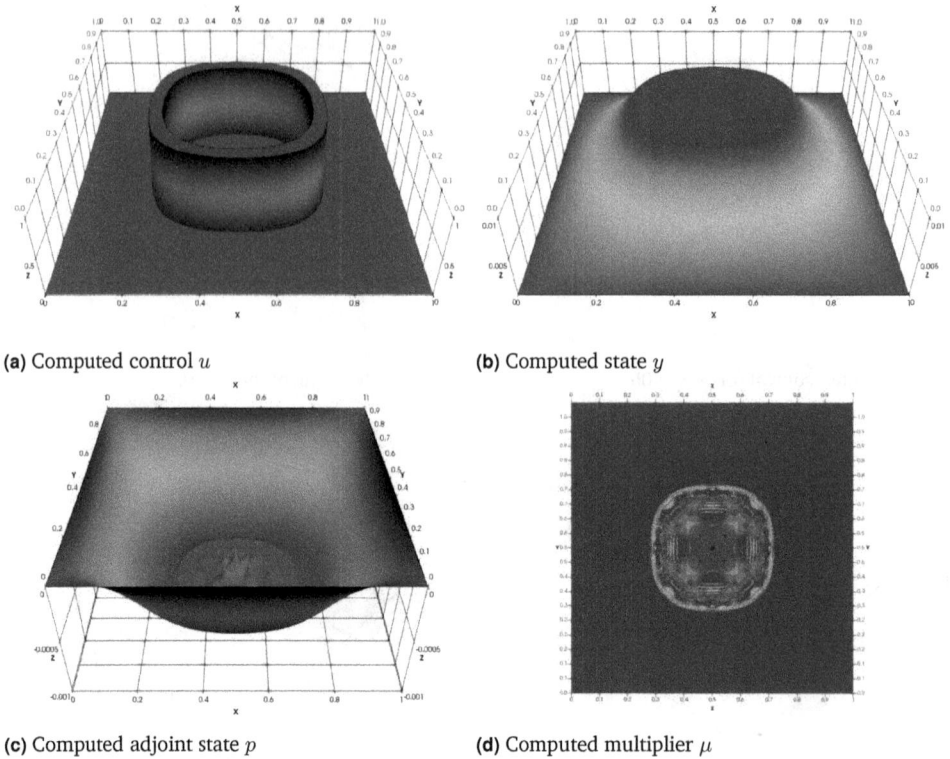

Figure 7.4: Computed results for Example 7.2.3. The range of μ is given by $\mu(x) \in [0, 40]$.

7.2.3 Bang-Bang-Off Solution on Unit Square

For the next example we set $\Omega = (0,1)^2$, $u_a = -1$, $u_b = 1$ and $\beta = 10^{-3}$. Furthermore $\tau = 0.8$, $\omega = 0.75$ and $\theta = 10$. Now define

$$\psi(x,y) := 0.01,$$

$$y_d(x,y) := \frac{1}{2\pi} \sin(\pi x) \sin(\pi y).$$

Note that here no exact solution is available. If the state constraint and the L^1-term are neglected, the exact solution is given by

$$\bar{y}(x,y) := y_d(x,y),$$
$$\bar{u}(x,y) := -\Delta y_d(x,y).$$

This example is taken from Subsection 5.2.2 and is an example of an optimal control problem where the desired state is reachable and the source condition $\bar{u} = S^* w$ with an element $w \in L^2(\Omega)$ is satisfied if the state constraints are not present.

β	final α	final ρ	successful steps	intermediate steps	not successful steps	average inner iterations
0.05	$2.38 \cdot 10^{-5}$	10^9	15	14	7	2.9
0.1	$2.38 \cdot 10^{-5}$	10^9	16	13	7	3.1
0.2	$3.17 \cdot 10^{-5}$	10^9	18	10	7	3.0
1	$5.6 \cdot 10^{-5}$	10^{10}	20	6	8	3.5

Table 7.1: Additional information for the computation of Example 7.2.2 for different β.

By adding the additional state constraints $y \leq 0.01$ we completely changed the structure of the solution.

We computed the solution on a regular triangular grid with $1.6 \cdot 10^5$ degrees of freedom, $\varepsilon = 10^{-6}$ and $\varepsilon_I = 5 \cdot 10^{-7}$. As starting values we set $\alpha = 0.1$ and $\rho = 100$. The algorithm stopped after 8 successful, 25 intermediate and 9 unsuccessful steps with the final values $\alpha = 0.1 \cdot 0.75^{33} \approx 7.5 \cdot 10^{-6}$ and $\rho = 100 \cdot 5^9 \approx 2.0 \cdot 10^8$. The computed results can be seen in Figure 7.4.

Part II

Nonlinear State Equation

CHAPTER 8

Tikhonov regularization

In all of the previous chapters we considered the minimization problem

$$\text{Minimize} \quad \frac{1}{2}\|Su - z\|_Y^2$$

$$\text{such that} \quad u_a \leq u \leq u_b \quad \text{a.e. in } \Omega,$$

with a linear operator S. In Chapter 5 and 6 we took S as the solution operator of the linear elliptic partial differential equation $-\Delta y = u$ with homogeneous Dirichlet boundary data. Although many physical problems can be modelled with this equation, in some cases a more sophisticated partial differential equation is needed. In this chapter we want to consider a semi-linear partial differential equation.

Optimal control of semi-linear partial differential equations has been intensively studied in the literature, see [7, 13, 15, 16, 19, 20, 93] and the references therein. As we show in this chapter, the results obtained for linear equations can be carried over using similar techniques while heavily relying on the second-order condition of Casas [13].

The work on regularization of optimal control problems is certainly connected to regularization of nonlinear inverse problems: If no control constraints are present, i.e., $U_{\text{ad}} = L^2(\Omega)$, the problem (P) described below is a heat source identification problem, which amounts to a nonlinear, ill-posed operator equation. Tikhonov regularization of nonlinear equations is studied, e.g., in the monograph [33]. Necessary conditions for convergence rates for non-linear problems can be found in [71]. Regularization of variational inequalities was studied in [63]. In some sense, our results generalize results from inverse problems theory: If no control constraints are present, our regularity conditions reduce to well-known source conditions.

First, we establish the model problem in Section 8.1 and state some preliminary results in Section 8.2, including a sufficient second-order condition. In this section we also extend the regularity assumption (ASC) to the non-linear case. This modified regularity assumption together with the second order condition is then used in Section 8.3 to derive regularization error estimates. It turns out that our regularity assumption is not only sufficient but also necessary for obtaining higher convergence rates. This fact is established in Section 8.4. Numerical results are presented in Section 8.5.

The results of this chapter can be found in condensed form in the publication [78].

8.1 Problem Setting

Throughout this section define the problem (P) to be the following optimal control problem

$$\text{Minimize} \quad J(u) = \frac{1}{2}\|y_u - y_d\|^2_{L^2(\Omega)}$$

$$\text{such that} \quad u_a \le u \le u_b \quad \text{a.e. in } \Omega, \qquad (P)$$

where y_u is the solution of the Dirichlet problem

$$Ay + f(x,y) = u \quad \text{in } \Omega,$$
$$y = 0 \quad \text{on } \partial\Omega. \qquad (8.1)$$

Here, $\Omega \subseteq \mathbb{R}^n$, $n \le 3$, is a bounded Lipschitz domain. The equation (8.1) is a semilinear elliptic equation with the linear elliptic operator A defined by

$$(Ay)(x) = \sum_{i,j=1}^{n} \partial_{x_j}[a_{ij}(x)\partial_{x_i} y(x)], \quad x \in \Omega.$$

Please note that the non-linearity f depends explicitly on the spatial variable x. We will suppress this dependency from now on, unless it is explicitly needed. The standing assumptions on the data of the problem will be made precise below.

Since the cost function J only implicitly depends on u through the solution y of the state equation, the control problem is in general not coercive with respect to u in suitable spaces. Optimal controls of (P) may exhibit a bang-bang structure where the control constraints are active on the whole domain, i.e., $\bar{u}(x) \in \{u_a, u_b\}$ almost everywhere. In addition, due to the nonlinear constraint (8.1) the resulting optimal control problem is non-convex. This makes the analysis and numerical solution of this problem challenging. To address this issue, we investigate the Tikhonov regularization of the problem given by

$$\text{Minimize} \quad J_\alpha(u) := \frac{1}{2}\|y - y_d\|^2_{L^2(\Omega)} + \frac{\alpha}{2}\|u\|^2_{L^2(\Omega)}$$

subject to the semilinear equation and the control constraints. Here, $\alpha > 0$ is the Tikhonov regularization parameter. We are interested in convergence of solutions or stationary points u_α of the regularized problems for $\alpha \to 0$. Under suitable conditions, we prove in Section 8.3 convergence rates of the type

$$\|u_\alpha - \bar{u}\|_{L^2(\Omega)} = \mathcal{O}\left(\alpha^{d/2}\right) \text{ for } \alpha \to 0,$$

see Theorem 8.3.8. This is the main result of the chapter, and it is a convergence rate result for regularization of optimal control problems subject to *nonlinear* partial differential equations. In addition, we also derive necessary conditions for convergence rates. As it turns out, a certain source condition is necessary to obtain convergence rates, see Section 8.4.

In the subsequent analysis, we will make use of the second-order conditions developed by Casas [13]. They require positive definiteness of the second-derivative J'' of the reduced cost functional with respect to solutions of linearized equations, see (SOSC) below. A second ingredient is a condition on the optimal control and adjoint state of the original problem, very similar to (ASC).

We used this condition to prove convergence rates for the iterative Bregman method in Chapter 3. However, there we considered a convex problem. Furthermore the regularity assumption (ASC) was used earlier for convex problems to prove convergence rates for Tikhonov regularization in [97]. In the following we want to continue this investigations and generalize these assumption to the non-convex case.

8.2 Assumptions and Preliminary Results

In the sequel, we will make use of the following assumptions, see [13]. To shorten our notation, we will denote the partial derivatives $\frac{\partial}{\partial y} f$ and $\frac{\partial^2}{\partial y^2} f$ by f' and f'', respectively.

(A1) We assume that $f : \Omega \times \mathbb{R} \to \mathbb{R}$ is a Carathéodory function, i.e. $f(\cdot, y)$ is measurable for all $y \in \mathbb{R}$ and $f(x, \cdot)$ is continuous for almost all $x \in \mathbb{R}$. Furthermore f is of class C^2 with respect to the second variable and satisfies $f(\cdot, 0) \in L^{\bar{p}}(\Omega)$, with $\bar{p} \geq \frac{n}{2}$, and
$$f'(x, y) \geq 0 \quad \forall y \in \mathbb{R}, \text{ for a.a. } x \in \Omega.$$

For all $M > 0$ there exists a constant $C_{f,M} > 0$ such that

$$|f'(x, y)| + |f''(x, y)| \leq C_{f,M} \quad \text{for a.a. } x \in \Omega \text{ and } |y| \leq M.$$

For every $M > 0$ and $\varepsilon > 0$ there exists $\delta > 0$, depending on M and ε, such that

$$|f''(x, y_2) - f''(x, y_1)| \leq \varepsilon$$

holds for all y_1, y_2 satisfying $|y_1|, |y_2| \leq M$, $|y_2 - y_1| \leq \delta$, and for a.a. $x \in \Omega$.

(A2) The coefficients of the operator A satisfy $a_{ij} \in C(\bar{\Omega})$. There exists some $\lambda_A > 0$ such that

$$\lambda_A |\zeta|^2 \leq \sum_{i,j=1}^{n} a_{ij}(x) \zeta_i \zeta_j \quad \forall \zeta \in \mathbb{R}^n, \quad \text{for a.a. } x \in \Omega.$$

(A3) We assume $y_d \in L^{\bar{s}}(\Omega)$, with $\bar{s} \geq \frac{n}{2}$. Moreover, $u_a, u_b \in L^\infty(\Omega)$ with $u_a \leq u_b$ a.e. on Ω.

Under these assumptions we can establish the following results. Existence and uniqueness of solutions of the state equations are well-known, see, e.g. [13, 15].

Theorem 8.2.1. *For every $u \in L^s(\Omega)$ with $s > \frac{n}{2}$, the state equation (8.1) has a unique solution $y_u \in H_0^1(\Omega) \cap C(\bar{\Omega})$. Moreover, the control-to-state mapping $S : L^s(\Omega) \to H_0^1(\Omega) \cap C(\bar{\Omega})$ is of class C^2 and globally Lipschitz continuous. For the special case $s = 2$ we obtain a $c > 0$ such that*

$$\|y_u\|_{H_0^1(\Omega)} + \|y_u\|_{C(\bar{\Omega})} \le c\|u\|_{L^2(\Omega)} \quad \forall u \in L^2(\Omega),$$

holds.

For convenience, let us introduce the space $Y := H_0^1(\Omega) \cap C(\bar{\Omega})$ endowed with the norm

$$\|y\|_Y := \|y\|_{H_0^1(\Omega)} + \|y\|_{C(\bar{\Omega})}$$

and the set of admissible functions

$$U_{\text{ad}} := \{u \in L^2(\Omega) : u_a \le u \le u_b\}.$$

Then Theorem 8.2.1 implies the existence of $M > 0$ such that

$$\|y_u\|_Y \le M \quad \forall u \in U_{\text{ad}}. \tag{8.2}$$

In addition, S maps weakly converging sequences to strongly converging sequences. The proof can be found in [15, Thm. 2.1].

Lemma 8.2.2. *Let (u_k) be a sequence in U_{ad} converging weakly in $L^2(\Omega)$ to u. Then, the associated sequence of states (y_k) converges strongly in Y to y_u.*

8.2.1 Existence of Solutions

The existence of solutions of the optimal control problem can be proved by classical arguments, see [93, Chapter 4].

Theorem 8.2.3. *Problem (P) has at least one solution $\bar{u}$ with an associated state $\bar{y} \in H_0^1(\Omega) \cap C(\bar{\Omega})$.*

The derivatives of the control-to-state map S can be characterized by the following systems. Let $u \in L^s(\Omega)$ be given with $y_u := S(u)$. Then for all $v \in L^2(\Omega)$, $z := S'(u)v$ is the unique weak solution of

$$
\begin{aligned}
Az + f'(y_u)z &= v && \text{in } \Omega, \\
z &= 0 && \text{on } \partial\Omega.
\end{aligned}
$$

In addition, let us introduce the adjoint state p_u associated to u as the unique weak solution of the adjoint equation

$$
\begin{aligned}
A^*p + f'(y_u)p &= y_u - y_d && \text{in } \Omega, \\
p &= 0 && \text{on } \partial\Omega.
\end{aligned}
$$

Following [13] we know that there exists a constant $c > 0$ such that $\|p_u\|_{L^\infty(\Omega)} \le c$ forall $u \in U_{ad}$. Furthermore by our assumptions on the semilinear equation, we know that $S'(u)v$ and $S'(u)^*v$ belong to $H_0^1(\Omega) \cap C(\bar\Omega)$ for all $v \in L^{\bar s}(\Omega)$, see [13]. Using these expressions, the derivatives of the cost functional J are given by the following lemma, see [13].

Lemma 8.2.4. *The functional* $J : L^2(\Omega) \to \mathbb{R}$ *is of class* C^2, *and the first and second derivative is given by*

$$J'(u)v = \int_\Omega p_u v \ dx,$$

$$J''(u)(v_1, v_2) = \int_\Omega (1 - f''(x, y_u)p_u) z_{v_1} z_{v_2} \ dx,$$

where we used the notation $z_{v_i} := S'(u)v_i$.

Let us recall the first-order necessary optimality conditions. We define for $\varepsilon > 0$

$$B_\varepsilon(\bar u) := \{u \in L^2(\Omega) : \|u - \bar u\|_{L^2(\Omega)} \le \varepsilon\}.$$

Theorem 8.2.5. *Let* $\bar u$ *be a local solution of problem* (P), *i.e. there exists an* $\varepsilon > 0$ *such that*

$$J(\bar u) \le J(u), \quad \forall u \in B_\varepsilon(\bar u) \cap U_{ad}.$$

Then there is $\bar y := S(\bar u) \in Y$ *and* $\bar p := p_{\bar u} \in H_0^1(\Omega)$ *such that the following system is satisfied:*

$$A\bar y + f(\bar y) = \bar u \quad \text{in } \Omega,$$
$$\bar y = 0 \quad \text{on } \partial\Omega, \tag{8.3}$$

$$A^*\bar p + f'(\bar y)\bar p = \bar y - y_d \quad \text{in } \Omega,$$
$$\bar p = 0 \quad \text{on } \partial\Omega, \tag{8.4}$$

$$J'(\bar u)(u - \bar u) \ge 0 \quad \forall u \in U_{ad}. \tag{8.5}$$

The variational inequality (8.5) implies the following relations between $\bar u$ and $\bar p$

$$\bar u(x) \begin{cases} = u_a(x) & \text{if } \bar p(x) > 0, \\ \in [u_a(x), u_b(x)] & \text{if } \bar p(x) = 0, \\ = u_b(x) & \text{if } \bar p(x) < 0. \end{cases} \tag{8.6}$$

Let us close this section with the following stability result regarding the solutions of the adjoint equations, see also [13, Lemma 2.5].

Lemma 8.2.6. *Let* $\bar u \in U_{ad}$ *be given with associated state* $\bar y$ *and adjoint state* $\bar p$. *Then there is a constant* $c > 0$ *such that for all* $u \in U_{ad}$ *it holds*

$$\|\bar p - p_u\|_Y \le c\|\bar y - y_u\|_{L^2(\Omega)}.$$

Proof. Let us denote $y := y_u$ and $p := p_u$. The functions $\bar{p}$ and p satisfy

$$A^*\bar{p} + f'(\bar{y})\bar{p} = \bar{y} - y_d,$$
$$A^*p + f'(y)p = y - y_d.$$

Then the difference $p - \bar{p}$ of the adjoint states satisfies

$$A^*(p - \bar{p}) + f'(y)(p - \bar{p}) = y - \bar{y} + (f'(\bar{y}) - f'(y))\bar{p}.$$

Due to the Lax-Milgram theorem, Stampacchia's estimates [90, Théorème 4.2] and [93], there is $c > 0$ such that

$$\|p - \bar{p}\|_Y \le c\|y - \bar{y} + (f'(\bar{y}) - f'(y))\bar{p}\|_{L^2(\Omega)}.$$

Since $\bar{p}$ is the solution of a linear elliptic equation with right-hand side in $L^2(\Omega)$, we know $\bar{p} \in L^\infty(\Omega)$. We now apply a Taylor expansion and obtain for almost all $x \in \Omega$ a $\zeta(x) \in \Omega$ such that

$$f'(x, \bar{y}(x)) - f'(x, y(x)) = f''(x, \zeta(x))(\bar{y}(x) - y(x)).$$

Using the assumptions on the function f we obtain by suppressing the spatial argument that

$$|f'(\bar{y}(x)) - f'(y(x))| \le C_{f,M}|\bar{y}(x) - y(x)|$$

holds for almost all $x \in \Omega$. Hence, we can estimate

$$\|(f'(\bar{y}) - f'(y))\bar{p}\|_{L^2(\Omega)} \le C_{f,M}\|\bar{y} - y\|_{L^2(\Omega)}\|\bar{p}\|_{L^\infty(\Omega)}.$$

The claim now follows. □

8.2.2 Second-Order Optimality Conditions

As already mentioned in the introduction of this chapter we know that the objective functional $\|Su - z\|_Y^2$ is non-convex in general. Hence, the first order conditions established in Theorem 8.2.5 are only necessary but not sufficient. In order to work with stationary points we have to introduce a sufficient second order condition.

Before we start let us consider the optimization problem

$$\min_{x \in C} g(x) \tag{8.7}$$

with $C \subset X$ convex, X a Banach space and $g : X \to \mathbb{R}$ Fréchet differentiable. Now let $\bar{x}$ be a local solution of (8.7), then it holds

$$g'(\bar{x})h \ge 0, \quad \forall h \in T_C(\bar{x}), \tag{8.8}$$

see [8]. Here $T_C(\bar{x})$ denotes the tangent cone. In our case we set $X = L^2(\Omega)$, $C = U_{\text{ad}}$, $x = u$ and $g(x) := J(u)$. A calculation reveals

$$T_{U_{\text{ad}}}(\bar{u}) = \left\{ v \in L^2(\Omega) : v(x) \begin{cases} \ge 0 & \text{if } \bar{u}(x) = u_a(x) \\ \le 0 & \text{if } \bar{u}(x) = u_b(x) \end{cases} \right\}. \tag{8.9}$$

Let $\bar{u} \in U_{\text{ad}}$. We now follow [8, 13] and define the cone of critical directions

$$C_{\bar{u}} := \left\{ v \in L^2(\Omega) : \; v(x) \begin{cases} \geq 0 & \text{if } \bar{u}(x) = u_a(x) \\ \leq 0 & \text{if } \bar{u}(x) = u_b(x) \end{cases}, \quad J'(\bar{u})v = 0 \right\}.$$

A critical direction has to satisfy two properties. First, it has to be an element of the tangent cone (8.9). Second it has to satisfy $J'(\bar{u})v = 0$. Note that by (8.8) it always holds $J'(\bar{u})v \geq 0$ if $\bar{u}$ is a minimizer of (P). Hence the critical cone represents those directions for which the first order condition (8.8) does not provide information about optimality of $\bar{u}$.

Using this cone the necessary second-order optimality conditions for a local minimum $\bar{u}$ of (P) can be written in the form, see [8], [17]

$$J''(\bar{u})v^2 \geq 0, \quad \forall v \in C_{\bar{u}}.$$

Note that we used that the set U_{ad} is polyhedric in $L^2(\Omega)$. Using Lemma 8.2.4 and (8.6) it is easy to see that

$$C_{\bar{u}} = \left\{ v \in L^2(\Omega) : \; v(x) \begin{cases} \geq 0 & \text{if } \bar{u}(x) = u_a(x) \\ \leq 0 & \text{if } \bar{u}(x) = u_b(x) \\ = 0 & \text{if } \bar{p}(x) \neq 0 \end{cases} \right\}$$

holds. Hence, the cone of critical direction coincides with the cone used in [13]. The main idea to prove the equality is to write

$$J'(\bar{u})v = \int_{\Omega} \bar{p}v \, dx = \int_{\{\bar{p}>0\}} \bar{p}v \, dx + \int_{\{\bar{p}<0\}} \bar{p}v \, dx.$$

One of our main goals is to handle solutions with a bang-bang structure, i.e. $\bar{p}(x) \neq 0$ almost everywhere in Ω. If $\bar{u}$ is a bang-bang function we obtain $C_{\bar{u}} = \{0\}$, hence the necessary second order condition does not provide any information. Thus, a sufficient second order condition cannot be posed on the cone $C_{\bar{u}}$. To overcome this one can increase the set $C_{\bar{u}}$.

Again, we follow [8, Section 3.3] and introduce the extended critical cone with $\eta \geq 0$

$$B_{\bar{u}}^{\eta} := \left\{ v \in L^2(\Omega) : \; v(x) \begin{cases} \geq 0 & \text{if } \bar{u}(x) = u_a(x) \\ \leq 0 & \text{if } \bar{u}(x) = u_b(x) \end{cases}, \quad J'(\bar{u})v \leq \eta \|v\|_{L^2(\Omega)} \right\}.$$

It is obvious that for $\eta = 0$ we obtain $B_{\bar{u}}^0 = C_{\bar{u}}$. However, following Casas [13], we define for $\tau > 0$ the set

$$C_{\bar{u}}^{\tau} = \left\{ v \in L^2(\Omega) : \; v(x) \begin{cases} \geq 0 & \text{if } \bar{u}(x) = u_a(x) \\ \leq 0 & \text{if } \bar{u}(x) = u_b(x) \\ = 0 & \text{if } |\bar{p}(x)| > \tau \end{cases} \right\}.$$

129

Here one can show $C_{\bar{u}}^\tau \subseteq B_{\bar{u}}^\eta$ with $\eta = \sqrt{|\Omega|}\tau$. Before we formulate the sufficient second order condition for (P) let us consider the regularized functional

$$J_\Lambda(u) := \frac{1}{2}\|Su - y_d\|_{L^2(\Omega)}^2 + \frac{\Lambda}{2}\|u\|_{L^2(\Omega)}^2$$

with $\Lambda > 0$ and its second-order conditions. The next result is taken from [13, Theorem 2.3]. Recall that $z_v := S'(\bar{u})v$.

Theorem 8.2.7. *Let $\bar{u} \in U_{ad}$ satisfy $J_\Lambda'(\bar{u})(u - \bar{u}) \geq 0$ for every $u \in U_{ad}$. Then the following are equivalent:*

1) $J_\Lambda''(\bar{u})v^2 > 0 \quad \forall v \in C_{\bar{u}} \setminus \{0\}$

2) $\exists \nu > 0$ and $\tau > 0$ s.t. $J_\Lambda''(\bar{u})v^2 \geq \nu\|v\|_{L^2(\Omega)}^2 \quad \forall v \in C_{\bar{u}}^\tau$

3) $\exists \nu > 0$ and $\tau > 0$ s.t. $J_\Lambda''(\bar{u})v^2 \geq \nu\|z_v\|_{L^2(\Omega)}^2 \quad \forall v \in C_{\bar{u}}^\tau$

The inclusions 2) $\to$ 3) $\to$ 1) are immediate and also hold for the case $\Lambda = 0$. It is known that condition 1) from Theorem 8.2.7 is not enough to guarantee local optimality in general, for an example see [13, Example 2.1] and condition 2) does not hold for the case $\Lambda = 0$, see [13, Section 2].

This motivates to use condition 3) as a second-order condition.

Assumption SOSC (Second order sufficient condition). *Let $\bar{u} \in U_{ad}$ be given. Assume that there exists $\delta > 0$ and $\tau > 0$ such that*

$$J''(\bar{u})(v, v) \geq \delta\|z_v\|_{L^2(\Omega)}^2 \quad \forall v \in C_{\bar{u}}^\tau,$$

where we used the notation $z_v = S'(\bar{u})v$.

This condition together with the first-order necessary condition implies local optimality, see [13, Corollary 2.8].

Theorem 8.2.9. *Let us assume that $\bar{u}$ is a feasible control for problem (P) satisfying the first order optimality conditions (8.3)–(8.5) and the second order condition (SOSC). Then, there exists $\varepsilon > 0$ such that*

$$J(\bar{u}) + \frac{\delta}{9}\|y_u - \bar{y}\|_{L^2(\Omega)}^2 \leq J(u) \quad \forall u \in B_\varepsilon(\bar{u}) \cap U_{ad}. \tag{8.10}$$

Let us remark that the growth condition (8.10) is different from the ones obtained in [8, 17]. There a growth condition of the form

$$J(\bar{u}) + c\|u - \bar{u}\|_{L^2(\Omega)}^2 \leq J(u) \quad \forall u \in B_\varepsilon(\bar{u}) \cap U_{ad}$$

is obtained. However, such a growth condition does not hold in our case, see [13].

Furthermore let us present some interesting results established in the recently published paper by Casas, Wachsmuth and Wachsmuth [20]. They showed that the structural assumption

$$\text{meas}(\{x \in \Omega : |\bar{p}(x)| \leq \varepsilon\}) \leq K\varepsilon \quad \forall \varepsilon > 0, \tag{8.11}$$

among with the assumptions (A1)-(A3) is enough for a stationary point $\bar{u}$ of (P) to satisfy the growth condition

$$J(\bar{u}) + c\|u - \bar{u}\|^2_{L^1(\Omega)} \le J(u) \quad \forall u \in U_{ad} \cap B^\infty_\varepsilon(\bar{u}) \tag{8.12}$$

for some $\varepsilon > 0$. Here $B^\infty_\varepsilon(\bar{u})$ is the ball with radius ε in $L^\infty(\Omega)$. Note that the structural assumption (8.11) is Assumption (ASC) with $A = \Omega$ and $\kappa = 1$. Hence this condition implies that $\bar{u}$ is a bang-bang control. It is quite interesting that no assumption on the second derivative of J is needed. However, one should keep in mind that this result is rather weak due to the L^∞-norm. If ε is too small and $\bar{u}$ is a bang-bang solution, then $U_{ad} \cap B^\infty_\varepsilon(\bar{u})$ contains no other bang-bang controls besides $\bar{u}$. Hence (8.12) is not suited to compare other bang-bang controls with $\bar{u}$.

In one of the main results in this paper the authors showed local optimality with respect to the L^1-norm under a second order condition.

Theorem 8.2.10. *Let $\bar{u} \in U_{ad}$ be a stationary point of (P) and assume that the structural assumption (8.11) is satisfied. Furthermore assume that there exists constants $\tau, c_1 > 0$ with $c_1 < (4\|u_b - u_a\|_{L^\infty(\Omega)}K)^{-1}$ such that*

$$J''(\bar{u})v^2 \ge -c_1\|v\|^2_{L^1(\Omega)} \quad \forall v \in C^\tau_{\bar{u}} \tag{8.13}$$

holds. Then there exists constant $\varepsilon, c_2 > 0$ such that

$$J(\bar{u}) + c\|u - \bar{u}\|^2_{L^1(\Omega)} \le J(u) \quad \forall u \in U_{ad} \cap B^1_\varepsilon(\bar{u}).$$

Here $B^1_\varepsilon(\bar{u})$ is the ball around $\bar{u}$ with radius ε in $L^1(\Omega)$.

The proof can be found in [20, Theorem 2.8]. Note that the second order condition allows negative curvature of the second derivative on critical directions. It is also remarkable, that the sufficient second order condition (8.13) can be equivalently rewritten using measures, see [20, Theorem 2.14].

8.2.3 Regularity Conditions

In order to derive regularization error estimates for the control we assume some regularity on $\bar{u}$. We say that $\bar{u}$ satisfies the assumption (ASC) if the following holds.

Assumption ASC (Active-Set Condition). *Let $\bar{u}$ be an element of U_{ad} satisfying (8.3)-(8.5). Assume that there exists a measurable set $I \subseteq \Omega$, a function $w \in L^2(\Omega)$, and positive constants κ, c such that the following holds:*

1. *(source condition)* $I \supset \{x \in \Omega : \bar{p}(x) = 0\}$ *and*

$$\bar{u} = P_{U_{ad}}(S'(\bar{u})^*w) \quad \text{a.e. in } I,$$

2. *(structure of active set)* $A := \Omega \setminus I$ *and for all $\varepsilon > 0$*

$$\text{meas}(\{x \in A : 0 < |\bar{p}(x)| < \varepsilon\}) \le c\varepsilon^\kappa.$$

This assumption differs from the regularity assumptions used in Chapter 3 just by the assumption on the regularity of the solution, i.e. $S'(\bar{u})^*w \in L^\infty(\Omega)$, which is guaranteed by the regularity of the partial differential equation.

This assumption is a combination of a source condition and a regularity assumption on the active sets. Similar regularity assumptions were used in, e.g., [77, 96, 97, 99] for problems with affine-linear control-to-state mapping S. Note that for the special case $A = \Omega$ the solution $\bar{u}$ is of bang-bang structure. Under this regularity assumption we can establish an improved first order necessary condition, see Lemma 3.4.18. The proofs coincide as it only uses the variational inequality (8.5).

Theorem 8.2.12. *Let $\bar{u}$ satisfy assumption (ASC), then there is a constant $c > 0$ such that it holds*

$$J'(\bar{u})(u - \bar{u}) \geq c\|u - \bar{u}\|_{L^1(A)}^{1+\frac{1}{\kappa}} \quad \forall u \in U_{\text{ad}}.$$

8.3 Convergence Results

In this section we want to combine the second order condition (SOSC) and the regularity assumption (ASC) to establish regularization error estimates.

8.3.1 Analysis of the Tikhonov Regularization

Let us introduce the Tikhonov regularized optimal control problem associated to (P). Let $\alpha > 0$ be given. Then the regularized problem reads

$$\text{Minimize} \quad J_\alpha(u) := \frac{1}{2}\|y_u - y_d\|_{L^2(\Omega)}^2 + \frac{\alpha}{2}\|u\|_{L^2(\Omega)}^2 \qquad (P_\alpha)$$
$$\text{such that} \quad u_a \leq u \leq u_b \quad \text{a.e. in } \Omega,$$

where y_u denotes again the solution of the semi-linear partial differential equation (8.1). Clearly, the regularized problem admits solutions.

At first, we want to show that weak limit points of global solutions $(u_\alpha)_\alpha$ for $\alpha \to 0$ are again global solutions of (P). In addition, we show that every strict local solution of (P) can be obtained as a limit of local solutions of (P_α). The results and the proofs are very similar to [18, Section 4], but since the proofs are short we present them here.

Lemma 8.3.1. *Let $(u_\alpha)_{\alpha>0}$ be a family of global solutions of (P_α) such that $u_\alpha \rightharpoonup u_0$ in $L^2(\Omega)$ for $\alpha \to 0$. Then u_0 is a global solution of (P). In addition, $u_\alpha \to u_0$ strongly in $L^2(\Omega)$. Moreover, the following identity holds*

$$\|u_0\|_{L^2(\Omega)} = \min\left\{\|u\|_{L^2(\Omega)} : u \text{ is a global solution of } (P)\right\}.$$

Proof. Let $u \in U_{\text{ad}}$ be given. Then it holds $J_\alpha(u_\alpha) \leq J_\alpha(u)$ for all $\alpha > 0$. The family $(u_\alpha)_\alpha$ is bounded in $L^\infty(\Omega)$. Then Lemma 8.2.2 implies

$$J_0(u_0) = \lim_{\alpha\to 0} J_0(u_\alpha) = \lim_{\alpha\to 0} J_\alpha(u_\alpha) \leq \lim_{\alpha\to 0} J_\alpha(u) = J_0(u).$$

Since $u \in U_{ad}$ was arbitrary, it follows that u_0 is a global solution of (P). Let us now prove the strong convergence $u_\alpha \to u_0$ in $L^2(\Omega)$. On the one hand, we have due to the weakly lower semicontinuity of the norm that

$$\|u_0\|_{L^2(\Omega)} \le \liminf_{\alpha \to 0} \|u_\alpha\|_{L^2(\Omega)} \le \limsup_{\alpha \to 0} \|u_\alpha\|_{L^2(\Omega)}.$$

On the other hand, using that u_0 is a global solution of (P), we obtain

$$
\begin{aligned}
J_0(u_\alpha) + \frac{\alpha}{2}\|u_\alpha\|_{L^2(\Omega)}^2 &= J_\alpha(u_\alpha) \le J_\alpha(u_0) = J_0(u_0) + \frac{\alpha}{2}\|u_0\|_{L^2(\Omega)}^2 \\
&\le J_0(u_\alpha) + \frac{\alpha}{2}\|u_0\|_{L^2(\Omega)}^2
\end{aligned}
\tag{8.14}
$$

which implies $\|u_\alpha\|_{L^2(\Omega)} \le \|u_0\|_{L^2(\Omega)}$ for all $\alpha > 0$. This shows $\|u_\alpha\|_{L^2(\Omega)} \to \|u_0\|_{L^2(\Omega)}$, and $u_\alpha \to u_0$ in $L^2(\Omega)$ follows. It remains to show the last equality. Let u be a global solution of (P). By replacing u_0 with u in (8.14) we obtain $\|u_\alpha\|_{L^2(\Omega)} \le \|u\|_{L^2(\Omega)}$ for all $\alpha > 0$. This shows

$$\|u_0\|_{L^2(\Omega)} = \lim_{\alpha \to 0} \|u_\alpha\|_{L^2(\Omega)} \le \|u\|_{L^2(\Omega)},$$

which finishes the proof. $\qquad\qquad\qquad\qquad\qquad\qquad\qquad\qquad\qquad\qquad\square$

This result shows that weak limit points of global solutions of (P_α) are global solutions of minimal norm of (P). Since this problem is non-convex in general, such minimal norm solutions may not be uniquely determined.

Theorem 8.3.2. *Let $\bar{u}$ be a strict local solution of (P). Then there exist $\bar{\alpha} > 0$ and a family $(u_\alpha)_{\alpha \in (0,\bar{\alpha})}$ of local solutions of (P_α) such that $u_\alpha \to \bar{u}$ in $L^2(\Omega)$ for $\alpha \to 0$.*

Proof. For $\rho > 0$ define the auxiliary feasible set $U_{ad,\rho} := U_{ad} \cap \{v \in L^2(\Omega) : \|v - \bar{u}\|_{L^2(\Omega)} \le \rho\}$. Let $\rho > 0$ be such that $\bar{u}$ is the unique global minimum of J_0 in the set $U_{ad,\rho}$. We investigate the following auxiliary problem:

$$\min J_\alpha(u) \quad \text{subject to } u \in U_{ad,\rho}.$$

For every $\alpha > 0$ let $u_{\rho,\alpha}$ be a global solution of this auxiliary problem. By construction, the family $(u_{\rho,\alpha})$ is uniformly bounded in $L^\infty(\Omega)$. Hence we find a sequence $\alpha_k \to 0$ such that $u_{\rho,\alpha_k} \rightharpoonup u_0$ in $L^2(\Omega)$. Arguing as in the proof of Lemma 8.3.1, it follows that u_0 is a global minimum of J_0 on $U_{ad,\rho}$ and $\|u_{\rho,\alpha} - u_0\|_{L^2(\Omega)} \to 0$. Consequently, we obtain $u_0 = \bar{u}$, and it holds $\lim_{\alpha \to 0} u_{\rho,\alpha} = \bar{u}$ strongly in $L^2(\Omega)$. This implies that there is $\bar{\alpha}$ such that $\|u_{\rho,\alpha} - \bar{u}\|_{L^2(\Omega)} < \rho$ for all $\alpha < \bar{\alpha}$. Thus, the controls $u_{\rho,\alpha}$ are local minima of J_α on U_{ad} for all $\alpha < \bar{\alpha}$. $\qquad\qquad\square$

Using the second-order optimality condition and the growth estimate of Theorem 8.2.9, we can establish the following a-priori error estimate for the states and adjoints. Analogous results were obtained in [97] for the case of a linear state equation.

Theorem 8.3.3. *Let $\bar{u}$ be a local solution of (P) satisfying (SOSC). Then there exists a family $(u_\alpha)_\alpha$ of local solutions of (P_α) such that for $\alpha \to 0$ the following estimates hold*

$$\|y_\alpha - \bar{y}\|_{L^2(\Omega)} = o(\sqrt{\alpha}), \quad \|p_\alpha - \bar{p}\|_{L^\infty(\Omega)} = o(\sqrt{\alpha}).$$

Proof. We follow the proof of Theorem 8.3.2 and obtain a $\rho > 0$ and a family $(u_\alpha)_\alpha$ of local solutions of (P_α) which are also global solutions to the auxiliary problem

$$\min J_\alpha(u) \quad \text{subject to } u \in U_{\text{ad},\rho}.$$

In addition, $J_\alpha(u_\alpha) \leq J_\alpha(\bar{u})$ holds for all $\alpha < \bar{\alpha}$, with $\bar{\alpha}$ as in the proof of Theorem 8.3.2. Using this inequality together with Theorem 8.2.9 we get

$$J_0(\bar{u}) + \frac{\delta}{9}\|y_\alpha - \bar{y}\|_{L^2(\Omega)}^2 + \frac{\alpha}{2}\|u_\alpha\|_{L^2(\Omega)}^2 \leq J_0(u_\alpha) + \frac{\alpha}{2}\|u_\alpha\|_{L^2(\Omega)}^2 = J_\alpha(u_\alpha)$$

$$\leq J_\alpha(\bar{u}) = J_0(\bar{u}) + \frac{\alpha}{2}\|\bar{u}\|_{L^2(\Omega)}^2.$$

This implies

$$\frac{\delta}{9}\|y_\alpha - \bar{y}\|_{L^2(\Omega)}^2 \leq \frac{\alpha}{2}\left(\|\bar{u}\|_{L^2(\Omega)}^2 - \|u_\alpha\|_{L^2(\Omega)}^2\right).$$

Using the strong convergence $u_\alpha \to \bar{u}$, we get

$$\lim_{\alpha \to 0} \frac{\|y_\alpha - \bar{y}\|_{L^2(\Omega)}}{\sqrt{\alpha}} = \lim_{\alpha \to 0} \frac{9}{2\delta}\sqrt{\|\bar{u}\|_{L^2(\Omega)}^2 - \|u_\alpha\|_{L^2(\Omega)}^2} \to 0,$$

which proves the first part of the claim. The second part follows directly from Lemma 8.2.6. $\qquad\square$

8.3.2 Convergence Rates

The results of Theorems 8.3.2 and 8.3.3 provide convergence results and a-priori rates. However, numerical computations reveal that the a-priori rates are suboptimal, see, e.g., the numerical example in Section 8.5. In addition, it is hard to guarantee that optimization algorithms deliver globally or locally optimal controls. Hence, we will assume in the subsequent analysis that only stationary points u_α of (P_α) are available. Recall that u_α is a stationary point if it satisfies

$$J'(u_\alpha)(u - u_\alpha) + (\alpha u_\alpha, u - u_\alpha) \geq 0 \quad \forall u \in U_{\text{ad}}.$$

Furthermore one observes that in many applications the optimal control $\bar{u}$ exhibits a bang-bang structure, as y_d is not reachable, i.e., there exists no feasible control $u \in U_{\text{ad}}$ such that $y_d = Su$. In this section we want to prove convergence rates under our regularity assumption (ASC), which is suitable for bang-bang solutions. The regularity assumption (ASC) was used in [77,96,97,99] to establish convergence rates for an affine-linear control-to-state mapping. First we need some technical results, which will be helpful later on. Recall that $z_v := S'(\bar{u})v$.

Lemma 8.3.4. *Let* $\bar{y} = S(\bar{u})$, $\bar{u} \in U_{\text{ad}}$ *be given. Then there are constants* $c > 0$ *and* $\varepsilon > 0$ *such that*

$$\|y_u - \bar{y}\|_{L^2(\Omega)} \leq c\|z_{u-\bar{u}}\|_{L^2(\Omega)}$$

holds for all y_u *with* $u \in U_{\text{ad}}$ *and* $\|y_u - \bar{y}\|_{L^2(\Omega)} \leq \varepsilon$.

Proof. This can be proven following the lines of [13, Corollary 2.8]. □

The following Lemma is an extension of [13, Lemma 2.7].

Lemma 8.3.5. *Let* $(u_\alpha)_\alpha$ *be a family of controls* $u_\alpha \in U_{\text{ad}}$ *such that* $u_\alpha \rightharpoonup \bar{u}$ *in* $L^2(\Omega)$ *for* $\alpha \to 0$. *Then for every* $\varepsilon > 0$ *there is* $\alpha_{\max} > 0$ *such that*

$$|J''(u_\alpha)v^2 - J''(\bar{u})v^2| \leq \varepsilon\|z_v\|^2_{L^2(\Omega)}$$

for all $\alpha \in (0, \alpha_{\max})$.

Proof. Let us denote the states and adjoints corresponding to u_α and $\bar{u}$ by y_α, p_α, and $\bar{y}$, $\bar{p}$, respectively. Due to Lemma 8.2.2 and 8.2.6 we obtain $y_\alpha \to \bar{y}$ and $p_\alpha \to \bar{p}$ in $L^\infty(\Omega)$. Let us define $z_{\alpha,v} := S'(u_\alpha)v$ and $z_v := S'(\bar{u})v$. According to Lemma 8.2.4 we can write

$$J''(u_\alpha)v^2 - J''(\bar{u})v^2 = \int_\Omega (1 - f''(y_\alpha)p_\alpha)z_{\alpha,v}^2 \, dx - \int_\Omega (1 - f''(\bar{y})\bar{p})z_v^2 \, dx$$

$$= \int_\Omega (f''(\bar{y})\bar{p} - f''(y_\alpha)p_\alpha)z_v^2 \, dx + \int_\Omega (1 - f''(y_\alpha)p_\alpha)(z_{\alpha,v}^2 - z_v^2) \, dx.$$

Here, the absolute value of the first integral can be made smaller than $\frac{\varepsilon}{2}\|z_v\|^2_{L^2(\Omega)}$ for α small enough due to $y_\alpha \to \bar{y}$ and $p_\alpha \to \bar{p}$ in $L^\infty(\Omega)$. It remains to study the second integral. Here we use the decomposition $z_{\alpha,v}^2 - z_v^2 = (z_{\alpha,v} + z_v)(z_{\alpha,v} - z_v)$. By definition $z_{\alpha,v} = S'(u_\alpha)v$ and $z_v = S'(\bar{u})v$ satisfy

$$Az_{\alpha,v} + f'(y_\alpha)z_{\alpha,v} = v,$$
$$Az_v + f'(\bar{y})z_v = v.$$

Hence, the difference $z_{\alpha,v} - z_v$ satisfies the differential equation

$$A(z_{\alpha,v} - z_v) + f'(y_\alpha)(z_{\alpha,v} - z_v) + (f'(y_\alpha) - f'(\bar{y}))z_v = 0.$$

Arguing as in Lemma 8.2.6 we find

$$\|z_{\alpha,v} - z_v\|_{L^2(\Omega)} \leq c\|f'(y_\alpha) - f'(\bar{y})\|_{L^\infty(\Omega)}\|z_v\|_{L^2(\Omega)}. \tag{8.15}$$

Note that the constant c is independent of y_α, which is a consequence of the non-negativity of f'. This can be seen by following the proof of the Lax-Milgram Theorem. With a similar argument used in Lemma 8.2.6 we obtain

$$\|f'(y_\alpha) - f'(\bar{y})\|_{L^\infty(\Omega)} \leq C_{f,M}\|y_\alpha - \bar{y}\|_{L^\infty}. \tag{8.16}$$

135

The right hand side of (8.16) can now be made arbitrarily small, since $y_\alpha \to \bar{y}$ in $L^\infty(\Omega)$. Using the inverse triangular inequality, the estimate (8.15) also implies the existence of $c > 0$ independent of α such that

$$\|z_{\alpha,v}\|_{L^2(\Omega)} \leq c\|z_v\|_{L^2(\Omega)}.$$

Due to our assumptions on the function f and the convergence $y_\alpha \to \bar{y}$ and $p_\alpha \to \bar{p}$ in $L^\infty(\Omega)$ we know that $1 - f''(y_\alpha)p_\alpha$ is uniformly bounded. Using the Cauchy-Schwarz inequality shows that the integral

$$\left| \int_\Omega (1 - f''(y_\alpha)p_\alpha)(z_{\alpha,v} + z_v)(z_{\alpha,v} - z_v) \, dx \right| \leq c\|z_{\alpha,v} + z_v\|_{L^2(\Omega)} \|z_{\alpha,v} - z_v\|_{L^2(\Omega)}$$

$$\leq c\|y_\alpha - \bar{y}\|_{L^\infty} \|z_v\|_{L^2(\Omega)}^2$$

can be made smaller than $\frac{\varepsilon}{2}\|z_v\|_{L^2(\Omega)}^2$ for α small enough. $\qquad\square$

In the subsequent analysis we need the following two results. Recall that a basic ingredient for the regularization error estimates for the iterative Bregman method in Chapter 3 was the estimate of $(u^\dagger, u^\dagger - u)_{L^2(\Omega)}$, see Lemma 3.4.19. It turns out that we need a similar result here.

Lemma 8.3.6. *Let $\bar{u}$ satisfy Assumption (ASC). Then it holds for all $u \in U_{\mathrm{ad}}$*

$$(\bar{u}, \bar{u} - u)_{L^2(\Omega)} \leq \|w\|_{L^2(\Omega)} \|z_{u-\bar{u}}\|_{L^2(\Omega)} + \|\bar{u} - S'(\bar{u})^*w\|_{L^\infty(A)} \|u - \bar{u}\|_{L^1(A)}.$$

Proof. Since U_{ad} is defined by pointwise inequalities, the projection onto U_{ad} can be taken pointwise. Let χ_U denote the characteristic function for the set U. Then the projection in item (ii) of Assumption (ASC) can be written as

$$\left(\chi_I(\bar{u} - S'(u)^*w), \ u - \bar{u} \right)_{L^2(\Omega)} \geq 0, \quad \forall u \in U_{\mathrm{ad}}.$$

We now compute

$$(\bar{u}, \bar{u} - u)_{L^2(\Omega)} = (\bar{u}, (\bar{u} - u) \,|_I)_{L^2(\Omega)} + (\bar{u}, (\bar{u} - u) \,|_A)_{L^2(\Omega)}$$
$$\leq (S'(\bar{u})^*w, (\bar{u} - u) \,|_I)_{L^2(\Omega)} + (\bar{u}, (\bar{u} - u) \,|_A)_{L^2(\Omega)}$$
$$= (w, S'(\bar{u})(\bar{u} - u))_{L^2(\Omega)} + (\bar{u} - S'(\bar{u})^*w, (\bar{u} - u) \,|_A)_{L^2(\Omega)}.$$

We now use that $S'(\bar{u})^*w \in L^\infty(\Omega)$ by our assumptions, which yields the result. $\qquad\square$

In the subsequent analysis we need the following result.

Lemma 8.3.7. *Let $\kappa > 0$ and $c_A > 0$ be given. Then there exists a constant $C \geq 0$ independent from α such that the following inequality holds*

$$\alpha\|\bar{u} - S'(\bar{u})^*w\|_{L^\infty(A)} \|u_\alpha - \bar{u}\|_{L^1(A)} \leq \frac{c_A}{2} \|u_\alpha - \bar{u}\|_{L^1(A)}^{1+\frac{1}{\kappa}} + C\alpha^{\kappa+1}.$$

Proof. We use Young's inequality to prove this result. Let $q, r > 0$ such that $q^{-1} + r^{-1} = 1$. Then, for every nonnegative $a, b \geq 0$ and positive $c > 0$ we obtain

$$ab = (cq)(a)\left(\frac{b}{cq}\right) \leq cq\left(\frac{a^q}{q} + \frac{b^r}{r(cq)^r}\right) = ca^q + \frac{cq}{r(cq)^r}b^r.$$

With the choices

$$q := 1 + \frac{1}{\kappa}, \qquad r := \kappa + 1,$$

and

$$a := \|u_\alpha - \bar{u}\|_{L^1(A)}, \qquad b := \alpha\|\bar{u} - S'(\bar{u})^*w\|_{L^\infty(A)}, \qquad c := \frac{c_A}{2},$$

the result is obtained with the constant

$$C := \frac{cq}{r(cq)^r}\|\bar{u} - S'(\bar{u})^*w\|_{L^\infty(A)}^{1+\kappa}.$$

$\square$

We now have everything at hand to establish convergence rates for the control. We want to point out, that we only need weak convergence of the sequence $(u_\alpha)_\alpha$.

Theorem 8.3.8. *Let $\bar{u}$ satisfy Assumption (ASC), and let the assumptions of Theorem 8.2.9 hold for $\bar{u}$. Let $(u_\alpha)_\alpha$ be a family of stationary points converging weakly in $L^2(\Omega)$ to $\bar{u}$ for $\alpha \to 0$. Then it holds with $d := \min(\kappa, 1)$ for $\alpha \to 0$*

$$\|z_{u_\alpha} - \bar{u}\|_{L^2(\Omega)} = \mathcal{O}\left(\alpha^{\frac{d+1}{2}}\right),$$

$$\|u_\alpha - \bar{u}\|_{L^1(A)} = \mathcal{O}\left(\alpha^{\frac{\kappa(d+1)}{\kappa+1}}\right),$$

$$\|u_\alpha - \bar{u}\|_{L^2(\Omega)} = \mathcal{O}\left(\alpha^{d/2}\right).$$

In the case $w = 0$ or $A = \Omega$, these convergence rates are obtained with $d := \kappa$.

Proof. Due to Theorem 8.2.9, $\bar{u}$ is a local minimum of (P). Since u_α is a stationary point for (P_α), we know

$$J'(u_\alpha)(u - u_\alpha) + \alpha(u_\alpha, u - u_\alpha)_{L^2(\Omega)} \geq 0 \quad \forall u \in U_{\mathrm{ad}}. \tag{8.17}$$

Due to the Assumption (ASC), Theorem 8.2.12 gives

$$J'(\bar{u})(u - \bar{u}) \geq c_A\|u - \bar{u}\|_{L^1(A)}^{1+\frac{1}{\kappa}} \quad \forall u \in U_{\mathrm{ad}}.$$

Using $\bar{u}$ and u_α as test functions in these inequalities and adding them, yields

$$c_A\|u_\alpha - \bar{u}\|_{L^1(A)}^{1+\frac{1}{\kappa}} + \alpha\|u_\alpha - \bar{u}\|_{L^2(\Omega)}^2 \leq \alpha(\bar{u}, \bar{u} - u_\alpha)_{L^2(\Omega)} + (J'(\bar{u}) - J'(u_\alpha))(u_\alpha - \bar{u}).$$

Using Lemma 8.3.6 and 8.3.7 we obtain

$$\alpha(\bar{u}, \bar{u} - u_\alpha)_{L^2(\Omega)} \leq \alpha\|w\|_{L^2(\Omega)}\|z_{u_\alpha - \bar{u}}\|_{L^2(\Omega)} + \alpha\|\bar{u} - S'(\bar{u})^*w\|_{L^\infty(A)}\|u_\alpha - \bar{u}\|_{L^1(A)}$$

$$\leq \alpha\|w\|_{L^2(\Omega)}\|z_{u_\alpha - \bar{u}}\|_{L^2(\Omega)} + \frac{c_A}{2}\|u_\alpha - \bar{u}\|_{L^1(A)}^{1+\frac{1}{\kappa}} + C\alpha^{\kappa+1},$$

with $C > 0$ independent of α. By Taylor expansion [85, Theorem 2.8], we obtain

$$(J'(\bar{u}) - J'(u_\alpha))(u_\alpha - \bar{u}) = -J''(\bar{u})(u_\alpha - \bar{u})^2 - \left(J''(\tilde{u}_\alpha) - J''(\bar{u})\right)(u_\alpha - \bar{u})^2,$$

with $\tilde{u}_\alpha := u_\alpha + \theta_\alpha(\bar{u} - u_\alpha)$ and $\theta_\alpha \in (0, 1)$.

Let us argue that $u_\alpha - \bar{u}$ is in the extended critical cone $C_{\bar{u}}^\tau$ for α small enough. Since $u_\alpha \rightharpoonup \bar{u}$ in $L^2(\Omega)$, it follows from Theorem 8.2.1, and Lemma 8.2.6 that $p_\alpha \to \bar{p}$ in $L^\infty(\Omega)$. Using the uniform boundedness of (u_α) in $L^\infty(\Omega)$, $\alpha \to 0$, and $p_\alpha \to \bar{p}$ in $L^\infty(\Omega)$ we obtain $|\alpha u_\alpha + p_\alpha| > \tau/2$ and $\text{sign}(\alpha u_\alpha + p_\alpha) = \text{sign}(\bar{p})$ for all α sufficiently small on the set, where $|\bar{p}| > \tau$ is satisfied. If we choose α small enough, then also $\frac{\tau}{2} > \alpha \max(\|u_a\|_{L^\infty}, \|u_b\|_{L^\infty})$ holds. The variational inequality (8.17) implies

$$u_\alpha = P_{U_{\text{ad}}}\left(-\frac{1}{\alpha}p_\alpha\right),$$

which yields $u_\alpha = \bar{u}$ on $|\bar{p}| > \tau$. Consequently, $u_\alpha - \bar{u} \in C_{\bar{u}}^\tau$ holds for all α sufficiently small. Hence, we can apply the second-order condition (SOSC) for $\bar{u}$ to obtain

$$J''(\bar{u})(u_\alpha - \bar{u})^2 \geq \delta \|z_{u_\alpha - \bar{u}}\|_{L^2(\Omega)}^2.$$

Let us now show that $\tilde{u}_\alpha \rightharpoonup \bar{u}$ in $L^2(\Omega)$. Let $g \in L^2(\Omega)$ and compute

$$(g, \tilde{u}_\alpha)_{L^2(\Omega)} = \underbrace{(g, u_\alpha)_{L^2(\Omega)}}_{\to (g, \bar{u})_{L^2(\Omega)}} + \underbrace{\theta_\alpha}_{0 < \theta_\alpha < 1} \underbrace{(g, \bar{u} - u_\alpha)}_{\to 0} \to (g, u_\alpha)_{L^2(\Omega)}.$$

Hence, we can apply Lemma 8.3.5 and find that

$$|J''(\tilde{u}_\alpha)v^2 - J''(\bar{u})v^2| \leq \frac{\delta}{4}\|z_v\|_{L^2(\Omega)}^2$$

for all α sufficiently small. Collecting the estimates above, we get

$$c_A\|u_\alpha - \bar{u}\|_{L^1(A)}^{1+\frac{1}{\kappa}} + \alpha\|u_\alpha - \bar{u}\|_{L^2(\Omega)}^2 \leq \alpha(\bar{u}, \bar{u} - u_\alpha)_{L^2(\Omega)} + (J'(u_\alpha) - J'(\bar{u}))(\bar{u} - u_\alpha)$$

$$\leq \alpha\|w\|_{L^2(\Omega)}\|z_{u_\alpha - \bar{u}}\|_{L^2(\Omega)} + \frac{c_A}{2}\|u_\alpha - \bar{u}\|_{L^1(A)}^{1+\frac{1}{\kappa}} + C\alpha^{\kappa+1}$$

$$- J''(\bar{u})(u_\alpha - \bar{u})^2 - (J''(\tilde{u}_\alpha) - J''(\bar{u}))(u_\alpha - \bar{u})^2$$

$$\leq \frac{\alpha^2\|w\|_{L^2(\Omega)}^2}{\delta} - \frac{\delta}{2}\|z_{u_\alpha - \bar{u}}\|_{L^2(\Omega)}^2 + \frac{c_A}{2}\|u_\alpha - \bar{u}\|_{L^1(A)}^{1+\frac{1}{\kappa}}.$$

This yields

$$\frac{\delta}{2}\|z_{u_\alpha - \bar{u}}\|_{L^2(\Omega)}^2 + \frac{c_A}{2}\|u_\alpha - \bar{u}\|_{L^1(A)}^{1+\frac{1}{\kappa}} + \alpha\|u_\alpha - \bar{u}\|_{L^2(\Omega)}^2 \leq \delta^{-1}\|w\|_{L^2(\Omega)}^2\alpha^2 + C\alpha^{\kappa+1},$$

which proves the claim. $\qquad\square$

Convergence rates for the state and adjoint state can now be easily obtained.

Corollary 8.3.9. *Let the assumptions of Theorem 8.3.8 hold for* $\bar{u}$. *Denote* $\bar{y}$ *the associated state and* $\bar{p}$ *the adjoint state. Then it holds for* $\alpha \to 0$

$$\|y_\alpha - \bar{y}\|_{L^2(\Omega)} = \mathcal{O}\left(\alpha^{\frac{d+1}{2}}\right), \quad \|p_\alpha - \bar{p}\|_{L^\infty(\Omega)} = \mathcal{O}\left(\alpha^{\frac{d+1}{2}}\right),$$

where d *is as in the statement of Theorem 8.3.8.*

Proof. By Theorem 8.3.8 we already know $\|z_{u_\alpha} - \bar{u}\|_{L^2(\Omega)} = \mathcal{O}\left(\alpha^{\frac{d+1}{2}}\right)$. Lemma 8.3.4 implies $\|y_\alpha - \bar{y}\|_{L^2(\Omega)} = \mathcal{O}\left(\alpha^{\frac{d+1}{2}}\right)$ for $\alpha \to 0$. Lemma 8.2.6 then proves the claim for the convergence of the adjoint states. □

Remark 8.3.10. *The convergence rates obtained in Theorem 8.3.8 and Corollary 8.3.9 resemble the rates obtained for the control of a linear partial differential equation, see [94, 95], which improved on the results of [97].*

8.4 Necessity of the Regularity Assumption

In this section we will show that the regularity assumption (ASC) is necessary to obtain the convergence rates provided by Theorem 8.3.8. In the case of a linear state equation, such results were obtained in [94, 95, 98]. As it turns out, these results can be transferred to the nonlinear case with suitable modifications.

Theorem 8.4.1. *Let $A \subset \Omega$ be a given set and define $I := \Omega \setminus A$. Let us assume that $\{x \in \Omega : \bar{p}(x) = 0\} \subset I$ holds. Furthermore assume that there exists a constant $\sigma > 0$ such that*

$$u_a(x) \leq -\sigma < 0 < \sigma \leq u_b(x) \quad \text{f.a.a. } x \in \Omega.$$

Let $(u_\alpha)_\alpha$ be a family of stationary points of (P_α) and $\bar{u}$ be a stationary point of (P). Suppose that

$$\|\bar{u} - u_\alpha\|_{L^1(A)} + \|\bar{p} - p_\alpha\|_{L^\infty(A)} = \mathcal{O}(\alpha^\kappa)$$

for some $\kappa > 1$ and all $\alpha > 0$ sufficiently small. Then there is $c > 0$ such that the relation

$$\text{meas}\left(\{x \in A : |\bar{p}(x)| \leq \varepsilon\}\right) \leq c\varepsilon^\kappa$$

is fulfilled for all $\varepsilon > 0$ sufficiently small.

Proof. The proof is analogous to that of the corresponding result [95, Thm. 22]. As this proof only uses the variational inequality (8.5), it can be transferred to our situation without modifications. □

Next, we will show that the source condition is satisfied on the set $\{x \in \Omega : \bar{p}(x) = 0\}$ if the convergence rate is sufficiently large. For a related result concerning the regularization of an ill-posed nonlinear operator equation we refer to [71].

Theorem 8.4.2. *Assume that $\bar{u}$ is a stationary point of (P). Let $(u_\alpha)_\alpha$ be a family of stationary points of (P_α) converging weakly to $\bar{u} \in U_{\mathrm{ad}}$ in $L^2(\Omega)$. Suppose the convergence rate $\|y_\alpha - \bar{y}\|_{L^2(\Omega)} = \mathcal{O}(\alpha)$ holds for $\alpha \to 0$. Then there exists a function $w \in L^2(\Omega)$ such that $\bar{u} = P_{U_{\mathrm{ad}}}(S'(\bar{u})^*w)$ holds pointwise almost everywhere on the set $K := \{x \in \Omega : \bar{p}(x) = 0\}$.*
If in addition $\|y_\alpha - \bar{y}\|_{L^2(\Omega)} = o(\alpha)$ holds, then $\bar{u}$ vanishes on K.

Proof. By assumptions, we know $u_\alpha \rightharpoonup \bar{u}$ in $L^2(\Omega)$, $\|y_\alpha - \bar{y}\|_{L^2(\Omega)} = \mathcal{O}(\alpha)$, and $\|p_\alpha - \bar{p}\|_Y = \mathcal{O}(\alpha)$, which is a consequence of Lemma 8.2.6.
Let $\hat{u} \in U_{\mathrm{ad}}$ be given with $\hat{u} = \bar{u}$ on $\Omega \setminus K$. This implies $(\bar{p}, \hat{u} - \bar{u})_{L^2(\Omega)} = 0$ and

$$0 \le (\bar{p}, u_\alpha - \bar{u})_{L^2(\Omega)} = (\bar{p}, u_\alpha - \hat{u})_{L^2(\Omega)}.$$

Testing the variational inequality (8.17) with $\hat{u}$ and adding it to the above leads to

$$(\alpha u_\alpha + p_\alpha - \bar{p}, \hat{u} - u_\alpha)_{L^2(\Omega)} \ge 0. \tag{8.18}$$

As in the predecessor works mentioned above, the idea of the proof is to divide this inequality by α and then to pass to the limit $\alpha \to 0$. Hence, we investigate the difference quotient $\frac{1}{\alpha}(p_\alpha - \bar{p})$. Using the defining equations of p_α and $\bar{p}$, we find that $p_\alpha - \bar{p}$ solves

$$\begin{aligned} A^*(p_\alpha - \bar{p}) + f'(\bar{y})(p_\alpha - \bar{p}) + (f'(y_\alpha) - f'(\bar{y}))p_\alpha &= y_\alpha - \bar{y} && \text{in } \Omega, \\ p_\alpha - \bar{p} &= 0 && \text{on } \partial\Omega. \end{aligned} \tag{8.19}$$

We use Taylor expansion in Banach spaces [44, Theorem 107] to obtain

$$f'(y_\alpha) - f'(\bar{y}) = \int_0^1 f''(\bar{y} + s(y_\alpha - \bar{y}))\, ds \, (y_\alpha - \bar{y}).$$

Since $y_\alpha - \bar{y}$ is uniformly bounded in $L^\infty(\Omega)$ by Theorem 8.2.1, the assumptions on f and the Lebesgue dominated convergence theorem imply $\int_0^1 f''(\bar{y} + s(y_\alpha - \bar{y}))\, ds \to f''(\bar{y})$ in $L^2(\Omega)$.
Let now $\dot{y}$ and $\dot{p}$ be subsequential weak limit points of $(\alpha^{-1}(y_\alpha - \bar{y}))$ and $(\alpha^{-1}(p_\alpha - \bar{p}))$ in $L^2(\Omega)$ and $H_0^1(\Omega)$, respectively. Dividing (8.19) by α and passing to the limit $\alpha \to 0$ yields

$$\begin{aligned} A^*\dot{p} + f'(\bar{y})\dot{p} + f''(\bar{y})\dot{y}\bar{p} &= \dot{y} && \text{in } \Omega, \\ \dot{p} &= 0 && \text{on } \partial\Omega. \end{aligned}$$

If needed we pick some suitable subsequences. This equation is to be understood in the weak sense, and the assumptions made on f and A allow us to pass the weak limit. Note that the assumptions imply $p_\alpha \to \bar{p}$ in $L^\infty(\Omega)$. This shows

$$\dot{p} = S'(\bar{u})^*(\dot{y} - f''(\bar{y})\dot{y}\bar{p}) =: S'(\bar{u})^*w$$

with $w := (1 - f''(\bar{y})\bar{p})\dot{y} \in L^2(\Omega)$. By the Rellich-Kondrachov-Theorem $H_0^1(\Omega)$ is compactly embedded in $L^2(\Omega)$, so $\dot{p}$ is a strong subsequential limit of $\alpha^{-1}(p_\alpha - \bar{p})$ in

$L^2(\Omega)$. Dividing the variational inequality (8.18) by α and passing to the limit $\alpha \to 0$ we find

$$0 \leq \limsup_{\alpha \to 0} (u_\alpha + \alpha^{-1}(p_\alpha - \bar{p}), \hat{u} - u_\alpha)_{L^2(\Omega)}$$

$$= \limsup_{\alpha \to 0} (-\|u_\alpha\|^2_{L^2(\Omega)}) + \lim_{\alpha \to 0} \left((u_\alpha, \hat{u})_{L^2(\Omega)} + \alpha^{-1}(p_\alpha - \bar{p}, \hat{u} - u_\alpha)_{L^2(\Omega)}\right)$$

$$\leq -\|\bar{u}\|^2_{L^2(\Omega)} + (\bar{u}, \hat{u})_{L^2(\Omega)} + (\dot{p}, \hat{u} - \bar{u})_{L^2(\Omega)}$$

$$= (\bar{u} + \dot{p}, \hat{u} - \bar{u})_{L^2(\Omega)}.$$

Note that we used the weakly lower semicontinuity of the norm here. Since $\hat{u} \in U_{\mathrm{ad}}$ was arbitrary with the restriction $\hat{u} = \bar{u}$ on $\Omega \setminus K$, this inequality implies

$$\chi_K \bar{u} = \chi_K P_{U_{\mathrm{ad}}}(-S'(\bar{u})^* w).$$

If in addition we have $\|y_\alpha - \bar{y}\|_{L^2(\Omega)} = o(\alpha)$, then we obtain $\|p_\alpha - \bar{p}\|_Y = o(\alpha)$. This implies that $\alpha^{-1}(p_\alpha - \bar{p})$ converges to zero in $L^\infty(\Omega)$. Passing to the limit in (8.18) gives $\chi_K \bar{u} = \chi_K P_{U_{\mathrm{ad}}}(0)$, hence $\bar{u} = P_{U_{\mathrm{ad}}}(0)$ holds almost everywhere on K. $\quad\square$

Remark 8.4.3. *Let us point out an interesting reformulation of the source condition in terms of the Lagrangian. To this end, let us introduce the Lagrange function to problem (P) by*

$$\mathcal{L}(y, u, p) := J(y) - \langle Ay + f(y) - u, p \rangle.$$

Then the result of the previous theorem can be written as: There exists $\dot{y} \in L^2(\Omega)$ such that

$$\chi_K \bar{u} = \chi_K P_{U_{\mathrm{ad}}}\left(- S'(\bar{u})^*(\mathcal{L}_{yy}(\bar{u}, \bar{y}, \bar{p})\dot{y})\right).$$

Here, $\mathcal{L}_{yy}$ denotes the partial derivative of second order of L with respect to y interpreted as a linear and continuous mapping from $L^2(\Omega)$ to $L^2(\Omega)$.
In case of a linear state equation, we obtain $L_{yy} = \mathrm{id}$. In this case, the theorem above reduces to the results obtained in [98].
In addition, the above results resemble results for nonlinear inverse problems from [71]. Under the assumptions $U_{\mathrm{ad}} = L^2(\Omega)$ and $\bar{y} = y_d$ (exact and attainable data), the source condition reduces to

$$\bar{u} = -S'(\bar{u})\dot{y}.$$

Here, we used that $\bar{y} = y_d$ implies $\bar{p} = 0$ and $L_{yy}(\bar{y}, \bar{u}, \bar{p}) = \mathrm{id}$.

8.5 Numerical Results

In this section we present a numerical example to support our theoretical results. We construct a bang-bang solution for the following optimal control problem:

$$\text{Minimize} \quad \frac{1}{2}\|y - y_d\|^2_{L^2(\Omega)} \tag{8.20}$$

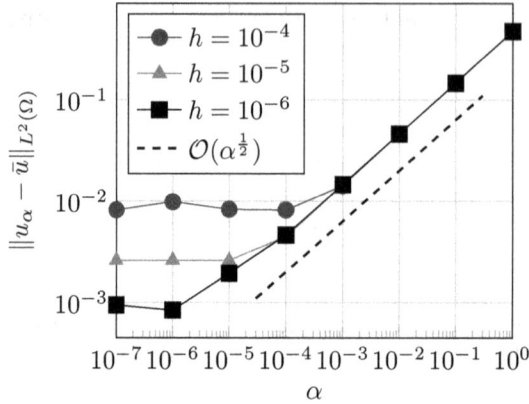

Figure 8.1: Error $\|u_\alpha - \bar{u}\|_{L^2(\Omega)}$ for $f(y) = \exp(y)$ in a double logarithmic plot for different values for h and α. For comparison we plotted a line with slope $\frac{1}{2}$.

subject to

$$-\Delta y + f(y) = u + e_\Omega \quad \text{in } \Omega$$
$$y = 0 \quad \text{on } \partial\Omega$$

and

$$-1 \leq u \leq 1 \quad \text{a.e. in } \Omega.$$

with $\Omega = (0,1)$. To solve the regularized optimal control problem numerically, we use dolfin-adjoint [35, 40]. We discretized control, state and adjoint state with linear finite elements. We make use of the adjoint equation

$$-\Delta\bar{p} + f'(\bar{y})\bar{p} = \bar{y} - y_d$$

and set

$$\bar{p}(x) := \sin(2\pi x),$$
$$\bar{u}(x) := -\text{sgn}(\bar{p}(x)),$$
$$\bar{y}(x) := \sin(\pi x),$$

and

$$e_\Omega(x) := -\bar{u}(x) - \Delta\bar{y}(x) + f(\bar{y}(x)),$$
$$y_d(x) := \bar{y}(x) + \Delta\bar{p}(x) - f'(\bar{y}(x))\bar{p}(x).$$

It is easy to check that $(\bar{u}, \bar{y}, \bar{p})$ is a solution of (8.20). Moreover, Assumption (ASC) is satisfied with $A = \Omega$ and $\kappa = 1$, see [28]. We expect to obtain the following convergence rate with respect to the L^2-norm:

$$\|u_\alpha - \bar{u}\|_{L^2(\Omega)} = \mathcal{O}\left(\alpha^{\frac{1}{2}}\right).$$

We define the non-linearity

$$f(x, y) := \exp(y).$$

The function f satisfies the assumptions made in (A1). We computed the problem on a equidistant subdivision of the interval Ω with different mesh sizes h and regularization parameter α. The results can be seen in Figure 8.1, where we plotted the error $\|u_\alpha - \bar{u}\|_{L^2(\Omega)}$ for solutions u_α of the discretized and regularized problem. As expected, the theoretical convergence order is very well resolved. In addition, we observe that for small α the discretization error dominates, which is an expected saturation effect.

CHAPTER 9

Extension to Sparse Control Problems

We now investigate sparse control problems given by

$$\text{Minimize} \quad \frac{1}{2}\|y_u - y_d\|_{L^2(\Omega)}^2 + \beta\|u\|_{L^1(\Omega)}$$

$$\text{such that} \quad u_a \le u \le u_b \quad \text{a.e. in } \Omega,$$

where $\beta > 0$ is a parameter. This is a non-smooth variant of the control problem presented in Chapter 8. Again we study the Tikhonov regularization and derive error estimates, see Section 9.5.

In recent years, there is a growing interest in sparse optimal control problems starting with [89], see also [13, 16]. Tikhonov regularization and its convergence was studied in [94, 95, 97] in connection with linear-quadratic optimal control problems.

In Section 9.1 we introduce the model problem and its Tikhonov regularization in Section 9.2. Sufficient second order conditions are presented in Section 9.3. Using the regularity assumption in Section 9.4 we establish convergence rates in Section 9.5.

9.1 Model Problem with Sparsity

In this section we consider the problem

$$\text{Minimize} \quad F(u) = J(u) + \beta j(u) = \frac{1}{2}\|y_u - y_d\|_{L^2(\Omega)}^2 + \beta\|u\|_{L^1(\Omega)} \qquad (P_\beta)$$

$$\text{such that} \quad u_a \le u \le u_b \quad \text{a.e. in } \Omega,$$

with $J(u) := \frac{1}{2}\|y_u - y_d\|_{L^2(\Omega)}^2$, $j(u) := \|u\|_{L^1(\Omega)}$, and $\beta > 0$. Again y_u is the solution of the nonlinear Dirichlet problem

$$Ay + f(x, y) = u \quad \text{in } \Omega,$$
$$y = 0 \quad \text{on } \partial\Omega.$$

We assume that all the assumption made in Section 8.2 are still valid throughout this section.

The motivation for the additional L^1-term in the cost functional F is the following. A solution $\bar{u}$ of (P_β) is sparse, i.e. large parts of $\bar{u}$ are identically zero. The larger β, the smaller the support of $\bar{u}$. One possible application of such a model is the optimal placement of controllers, since in many cases it is not desirable to control the system from the whole domain Ω. Starting with the pioneering work [89], such sparsity related control problems have been studied in, e.g., [97–99] for optimal control of linear partial differential equations and [13,16] for the optimal control of semi-linear equations.

In order to simplify the exposition, we assume $u_a(x) \le 0 \le u_b(x)$ almost everywhere in Ω. Our aim is to investigate so called bang-bang-off solutions, i.e., $\bar{u}(x) \in \{u_a(x), 0, u_b(x)\}$ almost everywhere in Ω. The necessary optimality conditions for problem (P_β) are given by:

$$A\bar{y} + f(\bar{y}) = \bar{u} \quad \text{in } \Omega,$$
$$\bar{y} = 0 \quad \text{on } \partial\Omega, \tag{9.1}$$

$$A^*\bar{p} + f'(\bar{y})\bar{p} = \bar{y} - y_d \quad \text{in } \Omega,$$
$$\bar{y} = 0 \quad \text{on } \partial\Omega, \tag{9.2}$$

$$\int_\Omega (\bar{p} + \beta\bar{\lambda})(u - \bar{u}) \, dx \ge 0 \quad \forall u \in U_{\text{ad}} \tag{9.3}$$

with $\bar{\lambda} \in \partial\|\bar{u}\|_{L^1(\Omega)}$. We refer to [13] for proofs. Similar to Subsection 6.2.2 we obtain for almost all $x \in \Omega$

$$\bar{\lambda} \begin{cases} = +1 & \text{if } \bar{u}(x) > 0, \\ = -1 & \text{if } \bar{u}(x) < 0, \\ \in [-1, +1] & \text{if } \bar{u}(x) = 0. \end{cases} \tag{9.4}$$

9.2 The Tikhonov Regularization

Again we consider the Tikhonov regularization of problem (P_β) given by

$$\text{Minimize} \quad F_\alpha(u) = \frac{1}{2}\|y_u - y_d\|_{L^2(\Omega)}^2 + \beta\|u\|_{L^1(\Omega)} + \frac{\alpha}{2}\|u\|_{L^2(\Omega)}^2 \tag{P_α}$$
$$\text{such that} \quad u_a \le u \le u_b \quad \text{a.e. in } \Omega.$$

The following convergence result can be proven similarly to the related result of Theorem 8.3.2.

Theorem 9.2.1. *Let $\bar{u}$ be a strict local solution of (P_β). Then there exist $\rho > 0$ and a family $(u_\alpha)_{\alpha \in (0,\bar{\alpha})}$ of local solutions of (P_α) such that $u_\alpha \to \bar{u}$ in $L^2(\Omega)$ for $\alpha \to 0$ and every u_α is a global minimum of F_α in $U_{\text{ad},\rho} := U_{\text{ad}} \cap \{v \in L^2(\Omega) : \|v - \bar{u}\|_{L^2(\Omega)} \le \rho\}$.*

9.3 Sufficient Second Order Conditions

Since j is not twice differentiable, we follow [13] and consider the modified extended critical cone defined by

$$\tilde{C}_u^\tau = \Big\{ v \in L^2(\Omega) : \; v(x) \geq 0 \text{ if } u(x) = u_a(x), \; v(x) \leq 0 \text{ if } u(x) = u_b(x),$$

$$\text{and } J'(u)v + \beta j'(u; v) \leq \tau \|z_v\|_{L^2(\Omega)} \Big\}.$$

The second order condition for the sparse control problem (P_β) reads as follows:

Assumption SOSC$_\beta$ (Second order sufficient condition). *Let $\bar{u} \in U_{\mathrm{ad}}$ be given. Assume that there exists $\delta > 0$ and $\tau > 0$ such that*

$$J''(\bar{u})v^2 \geq \delta \|z_v\|_{L^2(\Omega)}^2 \quad \forall v \in \tilde{C}_u^\tau.$$

This second order condition induces local quadratic growth of the cost functional. The next theorem is due to [13, Theorem 3.6].

Theorem 9.3.2. *Let us assume that $\bar{u}$ is a feasible control for problem (P_β) with state $\bar{y}$ and adjoint state $\bar{p}$ satisfying the first order optimality conditions (9.1)–(9.3) and the second order condition (SOSC$_\beta$). Then, there exists $\varepsilon > 0$ such that*

$$F(\bar{u}) + \frac{\delta}{5} \|z_{u-\bar{u}}\|_{L^2(\Omega)}^2 \leq F(u) \quad \forall u \in B_\varepsilon(\bar{u}) \cap U_{\mathrm{ad}}.$$

9.4 Regularity Assumption

The variational inequality (9.3) implies the following relations between $\bar{u}$ and $\bar{p}$

$$\bar{u}(x) \begin{cases} = u_a(x) & \text{if } \bar{p}(x) > \beta, \\ \in [u_a(x), 0] & \text{if } \bar{p}(x) = \beta, \\ = 0 & \text{if } |\bar{p}(x)| < \beta, \\ \in [0, u_b(x)] & \text{if } \bar{p}(x) = -\beta, \\ = u_b(x) & \text{if } \bar{p}(x) < -\beta, \end{cases} \tag{9.5}$$

see [13, Theorem 3.1] and [89, Section 2]. Hence, we have to modify the regularity assumption (ASC) to take the influence of the non-smooth term j into account, see also [97, 98].

Assumption ASC$_\beta$ (Active-Set Condition). *Let $\bar{u}$ be an element of U_{ad} satisfying (9.1)-(9.3), and assume that there exists a measurable set $I \subseteq \Omega$, a function $w \in L^2(\Omega)$, and positive constants κ, c such that the following holds*

1. *(source condition) $I \supset \{x \in \Omega : |\bar{p}(x)| = \beta\}$ and*

$$\bar{u} = P_{U_{\mathrm{ad}}}(S'(\bar{u})^* w) \quad \text{a.e. in } I,$$

2. *(structure of active set)* $A := \Omega \setminus I$ *and for all* $\varepsilon > 0$

$$\text{meas}\left(\{x \in A : \ 0 < \left|\left|\bar{p}(x)\right| - \beta\right| < \varepsilon\}\right) \le c\varepsilon^{\kappa}.$$

Note that if $\bar{u}$ satisfies this condition with $A = \Omega$ it exhibits a bang-bang-off structure, and the set $\{x \in \Omega : |\bar{p}(x)| = \beta\}$ is a set of measure zero. Again, we can establish an improved first order necessary condition.

Lemma 9.4.2. *Let $\bar{u}$ satisfy Assumption (ASC$_\beta$), then there is $c_A > 0$ such that*

$$J'(\bar{u})(u - \bar{u}) + \beta j'(\bar{u}; u - \bar{u}) \ge c_A \|u - \bar{u}\|_{L^1(A)}^{1+\frac{1}{\kappa}} \quad \forall u \in U_{\text{ad}}.$$

Proof. We start by computing the directional derivative of the objective functional. Let us define the function $v = u - \bar{u}$ for $u \in U_{\text{ad}}$ given. Now define, see [13, Proposition 3.3]

$$g(x) = \begin{cases} (\bar{p}(x) - \beta)v(x) & \text{if} \quad \bar{p}(x) > \beta, \\ (\bar{p}(x) + \beta)v(x) & \text{if} \quad \bar{p}(x) < -\beta, \\ \bar{p}(x)v(x) + \beta|v(x)| & \text{if} \quad |\bar{p}(x)| < \beta, \\ \beta(v(x) + |v(x)|) & \text{if} \quad \bar{p}(x) = \beta \text{ and } \bar{u}(x) = 0, \\ \beta(-v(x) + |v(x)|) & \text{if} \quad \bar{p}(x) = -\beta \text{ and } \bar{u}(x) = 0, \\ 0 & \text{else.} \end{cases}$$

Following [13] we know that $g(x) \ge 0$ almost everywhere in Ω and

$$J'(\bar{u})(u - \bar{u}) + \beta j'(\bar{u}; u - \bar{u}) = \int_\Omega g(x) \, \mathrm{d}x$$

holds. Using (9.5) and (9.4) we can now compute

$$J'(\bar{u})(u - \bar{u}) + \beta j'(\bar{u}; u - \bar{u}) \ge \int_{\{|\bar{p}| > \beta\}} (\bar{p} + \beta\bar{\lambda})(u - \bar{u}) \, \mathrm{d}x + \int_{\{|\bar{p}| < \beta\}} \bar{p}(u - \bar{u}) + \beta|u - \bar{u}| \, \mathrm{d}x,$$

where we used the abbreviation $\{|\bar{p}| > \beta\} := \{x \in \Omega : |\bar{p}(x)| > \beta\}$ and similar for $\{|\bar{p}| < \beta\}$.

Let $\varepsilon > 0$ be given. Then we have the following inclusion

$$\{|\bar{p}| > \beta\} \supseteq \{|\bar{p}| > \beta + \varepsilon\} =: M. \tag{9.6}$$

We now split the set $\{|\bar{p}| > \beta\}$ into the following two disjoint sets

$$M_1 := \{|\bar{p}| > \beta + \varepsilon, \ \bar{p} < -\beta\},$$
$$M_2 := \{|\bar{p}| > \beta + \varepsilon, \ \bar{p} > \beta\},$$
$$M = M_1 \cup M_2.$$

Note that on M_1 we have $|\bar{p}| - \beta = -\bar{p} - \beta > \varepsilon$. Since $\varepsilon > 0$ this implies $|\bar{p} + \beta| > \varepsilon$. With a similar calculation on M_2 we obtain the following inequalities

$$
\begin{aligned}
|\bar{p} + \beta| &> \varepsilon \quad \text{on } M_1, \\
|\bar{p} - \beta| &> \varepsilon \quad \text{on } M_2.
\end{aligned}
\tag{9.7}
$$

We now use (9.6), (9.7), and the non-negativity of g to compute

$$
\begin{aligned}
\int_{\{|\bar{p}|>\beta\}} (\bar{p} + \beta\bar{\lambda})(u - \bar{u}) \ \mathrm{d}x &\geq \int_M (\bar{p} + \beta\bar{\lambda})(u - \bar{u}) \ \mathrm{d}x \\
&= \int_{M_1} (\bar{p} + \beta\bar{\lambda})(u - \bar{u}) \ \mathrm{d}x + \int_{M_2} (\bar{p} + \beta\bar{\lambda})(u - \bar{u}) \ \mathrm{d}x \\
&= \int_{M_1} \overbrace{|\bar{p} + \beta|}^{\geq \varepsilon} |u - \bar{u}| \ \mathrm{d}x + \int_{M_2} \overbrace{|\bar{p} - \beta|}^{\geq \varepsilon} |u - \bar{u}| \ \mathrm{d}x \\
&\geq \varepsilon \int_{M_1} |u - \bar{u}| \ \mathrm{d}x + \varepsilon \int_{M_2} |u - \bar{u}| \ \mathrm{d}x \\
&= \varepsilon \int_M |u - \bar{u}| \ \mathrm{d}x.
\end{aligned}
$$

It remains to estimate the second term appearing in the directional derivative. Again we use an inclusion of the following form

$$
\{|\bar{p}| < \beta\} \supseteq \{|\bar{p}| < \beta - \varepsilon\} =: N,
\tag{9.8}
$$

and split N into the following two disjoint sets

$$
\begin{aligned}
N_1 &:= \{|\bar{p}| < \beta - \varepsilon, \ \bar{p} \geq 0\}, \\
N_2 &:= \{|\bar{p}| < \beta - \varepsilon, \ \bar{p} < 0\}, \\
N &= N_1 \cup N_2.
\end{aligned}
$$

Similar to (9.7) we can prove the following inequalities

$$
\begin{aligned}
-\bar{p} &\geq \varepsilon - \beta \quad \text{on } N_1, \\
\bar{p} &\geq \varepsilon - \beta \quad \text{on } N_2.
\end{aligned}
\tag{9.9}
$$

This now reveals with (9.8), (9.9) and the non-negativity of g

$$\int_{\{|\bar{p}|<\beta\}} \bar{p}(u-\bar{u}) + \beta|u-\bar{u}| \ dx \geq \int_N \bar{p}(u-\bar{u}) + \beta|u-\bar{u}| \ dx$$

$$= \int_{N_1} \bar{p}(u-\bar{u}) + \beta|u-\bar{u}| \ dx + \int_{N_2} \bar{p}(u-\bar{u}) + \beta|u-\bar{u}| \ dx$$

$$\geq \int_{N_1} \overbrace{(-\bar{p})}^{\geq \varepsilon-\beta} |u-\bar{u}| + \beta|u-\bar{u}| \ dx + \int_{N_2} \overbrace{\bar{p}}^{\geq \varepsilon-\beta} |u-\bar{u}| + \beta|u-\bar{u}| \ dx$$

$$\geq \int_{N_1} (\varepsilon-\beta)|u-\bar{u}| + \beta|u-\bar{u}| \ dx + \int_{N_2} (\varepsilon-\beta)|u-\bar{u}| + \beta|u-\bar{u}| \ dx$$

$$= \varepsilon \int_{N_1} |u-\bar{u}| \ dx + \varepsilon \int_{N_2} |u-\bar{u}| \ dx$$

$$= \varepsilon \int_N |u-\bar{u}| \ dx.$$

Let us define

$$A_\varepsilon := \{x \in A : \ ||\bar{p}(x)| - \beta| \geq \varepsilon\} = M \cup N.$$

Then the above computations yield

$$J'(\bar{u})(u-\bar{u}) + \beta j'(\bar{u}; u-\bar{u}) \geq \varepsilon\|u-\bar{u}\|_{L^1(A_\varepsilon)}.$$

Let us note that Assumption (ASC$_\beta$) implies meas $(A \setminus A_\varepsilon) \leq c\varepsilon^\kappa$. Now, putting everything together, we obtain using the regularity assumption on the active set

$$J'(\bar{u})(u-\bar{u}) + \beta j'(\bar{u}; u-\bar{u}) \geq \varepsilon\|u-\bar{u}\|_{L^1(A_\varepsilon)}$$
$$= \varepsilon\|u-\bar{u}\|_{L^1(A)} - \varepsilon\|u-\bar{u}\|_{L^1(A\setminus A_\varepsilon)}$$
$$\geq \varepsilon\|u-\bar{u}\|_{L^1(A)} - \varepsilon\|u-\bar{u}\|_{L^\infty(\Omega)} \text{ meas } (A \setminus A_\varepsilon)$$
$$\geq \varepsilon\|u-\bar{u}\|_{L^1(A)} - c\varepsilon^{\kappa+1}.$$

Here $c > 1$ is a constant independent from u. Setting $\varepsilon = c^{-\frac{2}{\kappa}}\|u-\bar{u}\|_{L^1(A)}^{\frac{1}{\kappa}}$ proves the claim. $\qquad\square$

9.5 Convergence Rates

We are now in the position to prove convergence rates. The proof mainly follows the lines of Theorem 8.3.8.

Theorem 9.5.1. *Let $\bar{u}$ satisfy Assumption (ASC$_\beta$) and let the assumptions of Theorem 9.3.2 hold for $\bar{u}$. Let $(u_\alpha)_\alpha$ be a family of stationary points converging weakly in $L^2(\Omega)$ to $\bar{u}$. Then it holds with $d = \min(\kappa, 1)$ for $\alpha \to 0$ sufficiently small*

$$\|z_{u_\alpha} - \bar{u}\|_{L^2(\Omega)} = \mathcal{O}\left(\alpha^{\frac{d+1}{2}}\right),$$

$$\|u_\alpha - \bar{u}\|_{L^1(A)} = \mathcal{O}\left(\alpha^{\frac{\kappa(d+1)}{\kappa+1}}\right),$$

$$\|u_\alpha - \bar{u}\|_{L^2(\Omega)} = \mathcal{O}\left(\alpha^{d/2}\right).$$

In the case $w = 0$ or $A = \Omega$, these convergence rates are obtained with $d := \kappa$.

Proof. Due to Theorem 9.3.2, $\bar{u}$ is a strict local minimum. We split the proof in two parts and consider the two cases $u_\alpha - \bar{u} \in \tilde{C}_{\bar{u}}^\tau$ and $u_\alpha - \bar{u} \notin \tilde{C}_{\bar{u}}^\tau$.
Let us start with the case $u_\alpha - \bar{u} \in \tilde{C}_{\bar{u}}^\tau$. The optimality conditions for u_α and Lemma 9.4.2 reveal

$$(p_\alpha + \alpha u_\alpha, u - u_\alpha)_{L^2(\Omega)} + \beta j'(u_\alpha; u - u_\alpha) \geq 0 \quad \forall u \in U_{\text{ad}}, \tag{9.10}$$

$$(\bar{p}, u - \bar{u})_{L^2(\Omega)} + \beta j'(\bar{u}; u - \bar{u}) \geq c_A \|u - \bar{u}\|_{L^1(A)}^{1+\frac{1}{\kappa}} \quad \forall u \in U_{\text{ad}}. \tag{9.11}$$

Note that j is a convex function, hence we have the inequality

$$j'(x; y - x) \leq j(y) - j(x),$$

leading to

$$j'(u_\alpha; \bar{u} - u_\alpha) + j'(\bar{u}; u_\alpha - \bar{u}) \leq 0.$$

Testing (9.10) and (9.11) with $\bar{u}$ and u_α, respectively, we obtain

$$c_A \|u_\alpha - \bar{u}\|_{L^1(A)}^{1+\frac{1}{\kappa}} + \alpha \|u_\alpha - \bar{u}\|_{L^2(\Omega)}^2$$
$$\leq \alpha(\bar{u}, \bar{u} - u_\alpha)_{L^2(\Omega)} + (p_\alpha - \bar{p}, \bar{u} - u_\alpha)_{L^2(\Omega)}$$
$$+ \beta(j'(u_\alpha; \bar{u} - u_\alpha) + j'(\bar{u}; u_\alpha - \bar{u}))$$
$$\leq \alpha(\bar{u}, \bar{u} - u_\alpha)_{L^2(\Omega)} + (J'(u_\alpha) - J'(\bar{u}))(\bar{u} - u_\alpha).$$

As the regularity assumptions (ASC) and (ASC$_\beta$) only differ in item (ii), Lemma 8.3.6 is applicable here as well, which gives with Lemma 8.3.7

$$\alpha(\bar{u}, \bar{u} - u_\alpha)_{L^2(\Omega)} \leq \alpha \|w\|_{L^2(\Omega)} \|z_{u_\alpha} - \bar{u}\|_{L^2(\Omega)} + \alpha \|\bar{u} - S'(\bar{u})^* w\|_{L^\infty(A)} \|u_\alpha - \bar{u}\|_{L^1(A)}$$
$$\leq \alpha \|w\|_{L^2(\Omega)} \|z_{u_\alpha} - \bar{u}\|_{L^2(\Omega)} + \frac{c_A}{2} \|u_\alpha - \bar{u}\|_{L^1(A)}^{1+\frac{1}{\kappa}} + C\alpha^{\kappa+1},$$

with $C > 0$ independent of α. By Taylor expansion, we obtain

$$(J'(\bar{u}) - J'(u_\alpha))(u_\alpha - \bar{u}) = -J''(\bar{u})(u_\alpha - \bar{u})^2 - (J''(\tilde{u}_\alpha) - J''(\bar{u}))(u_\alpha - \bar{u})^2,$$

with $\tilde{u}_\alpha := u_\alpha + \theta_\alpha(\bar{u} - u_\alpha)$ and $\theta_\alpha \in (0,1)$. Since $u_\alpha - \bar{u} \in \tilde{C}_{\bar{u}}^\tau$ we can apply the second-order condition on $\bar{u}$ to obtain

$$J''(\bar{u})(u_\alpha - \bar{u})^2 \geq \delta \|z_{u_\alpha} - \bar{u}\|_{L^2(\Omega)}^2.$$

Similar to the proof of Theorem 8.3.8 we get $\tilde{u}_\alpha \rightharpoonup \bar{u}$ and hence by Lemma 8.3.5, we find that

$$|J''(\tilde{u}_\alpha)v^2 - J''(\bar{u})v^2| \leq \frac{\delta}{4}\|z_v\|_{L^2(\Omega)}^2$$

for all α sufficiently small. Altogether, we obtain

$$
\begin{aligned}
c_A\|u_\alpha - \bar{u}\|_{L^1(A)}^{1+\frac{1}{\kappa}} + \alpha\|u_\alpha - \bar{u}\|_{L^2(\Omega)}^2 &\leq \alpha(\bar{u}, \bar{u} - u_\alpha)_{L^2(\Omega)} + (J'(u_\alpha) - J'(\bar{u}))(\bar{u} - u_\alpha) \\
&\leq \alpha\|w\|_{L^2(\Omega)}\|z_{u_\alpha - \bar{u}}\|_{L^2(\Omega)} + \frac{c_A}{2}\|u_\alpha - \bar{u}\|_{L^1(A)}^{1+\frac{1}{\kappa}} + C\alpha^{\kappa+1} \\
&\quad - J''(\bar{u})(u_\alpha - \bar{u})^2 - (J''(\tilde{u}_\alpha) - J''(\bar{u}))(u_\alpha - \bar{u})^2 \\
&\leq \alpha^2\delta^{-1}\|w\|_{L^2(\Omega)}^2 - \frac{\delta}{2}\|z_{u_\alpha - \bar{u}}\|_{L^2(\Omega)}^2 + \frac{c_A}{2}\|u_\alpha - \bar{u}\|_{L^1(A)}^{1+\frac{1}{\kappa}}.
\end{aligned}
$$

This yields

$$\frac{\delta}{2}\|z_{u_\alpha - \bar{u}}\|_{L^2(\Omega)}^2 + \frac{c_A}{2}\|u_\alpha - \bar{u}\|_{L^1(A)}^{1+\frac{1}{\kappa}} + \alpha\|u_\alpha - \bar{u}\|_{L^2(\Omega)}^2 \leq \delta^{-1}\|w\|_{L^2(\Omega)}^2\alpha^2 + C\alpha^{\kappa+1},$$

which implies the existence of $C > 0$ such that

$$\|z_{u_\alpha - \bar{u}}\|_{L^2(\Omega)}^2 + \|u_\alpha - \bar{u}\|_{L^1(A)}^{1+\frac{1}{\kappa}} + \alpha\|u_\alpha - \bar{u}\|_{L^2(\Omega)} \leq C(\alpha^{\kappa+1} + \alpha^2)$$

holds for all α sufficiently small.

Now for the second case $u_\alpha - \bar{u} \notin \tilde{C}_{\bar{u}}^\tau$. By definition of the extended critical cone, we know

$$J'(\bar{u})(u_\alpha - \bar{u}) + \beta j'(\bar{u}; u_\alpha - \bar{u}) > \tau\|z_{u_\alpha - \bar{u}}\|_{L^2(\Omega)}.$$

Combining this with Lemma 9.4.2 yields

$$J'(\bar{u})(u_\alpha - \bar{u}) + \beta j'(\bar{u}; u_\alpha - \bar{u}) > \frac{\tau}{2}\|z_{u_\alpha - \bar{u}}\|_{L^2(\Omega)} + \frac{c_A}{2}\|u_\alpha - \bar{u}\|_{L^1(A)}^{1+\frac{1}{\kappa}}.$$

Using the expansion

$$(J'(\bar{u}) - J'(u_\alpha))(u_\alpha - \bar{u}) = -J''(\tilde{u}_\alpha)(u_\alpha - \bar{u})^2$$

with $\tilde{u}_\alpha := u_\alpha + \theta_\alpha(\bar{u} - u_\alpha)$ and $\theta_\alpha \in (0,1)$, we get similarly as in the first part of the proof

$$
\begin{aligned}
&\frac{\tau}{2}\|z_{u_\alpha - \bar{u}}\|_{L^2(\Omega)} + \frac{c_A}{2}\|u_\alpha - \bar{u}\|_{L^1(A)}^{1+\frac{1}{\kappa}} + \alpha\|u_\alpha - \bar{u}\|_{L^2(\Omega)}^2 \\
&\leq \alpha\|w\|_{L^2(\Omega)}\|z_{u_\alpha - \bar{u}}\|_{L^2(\Omega)} + \frac{c_A}{4}\|u_\alpha - \bar{u}\|_{L^1(A)}^{1+\frac{1}{\kappa}} + C\alpha^{\kappa+1} - J''(\tilde{u}_\alpha)(u_\alpha - \bar{u})^2.
\end{aligned}
$$

Using the structure of J'' given in Lemma 8.2.4 we obtain

$$|J''(\bar{u})(u_\alpha - \bar{u})^2| = \left| \int_\Omega (1 - f''(\bar{y})\bar{p})|z_{u_\alpha - \bar{u}}|^2 \, dx \right| \leq c\|z_{u_\alpha - \bar{u}}\|_{L^2(\Omega)}^2.$$

Here we used the boundedness of $\bar{y}$ and $\bar{p}$ in $L^\infty(\Omega)$ as well as the assumption on f. By Lemma 8.3.5, we obtain

$$|(J''(\tilde{u}_\alpha) - J''(\bar{u})(u_\alpha - \bar{u})^2| \le c\|z_{u_\alpha - \bar{u}}\|^2_{L^2(\Omega)}.$$

Note that we used again the convergence $\tilde{u}_\alpha \rightharpoonup \bar{u}$. Combining both results yield

$$|J''(\tilde{u}_\alpha)(u_\alpha - \bar{u})^2| \le |J''(\bar{u})(u_\alpha - \bar{u})^2| + |(J''(\tilde{u}_\alpha) - J''(\bar{u})(u_\alpha - \bar{u})^2|$$
$$\le c\|z_{u_\alpha - \bar{u}}\|^2_{L^2(\Omega)}$$

for all α sufficiently small. Hence, it holds

$$\frac{\tau}{2}\|z_{u_\alpha - \bar{u}}\|_{L^2(\Omega)} + \frac{c_A}{4}\|u_\alpha - \bar{u}\|^{1+\frac{1}{\kappa}}_{L^1(A)} + \alpha\|u_\alpha - \bar{u}\|^2_{L^2(\Omega)}$$
$$\le \alpha^2\|w\|^2_{L^2(\Omega)} + C\alpha^{\kappa+1} + c\|z_{u_\alpha - \bar{u}}\|^2_{L^2(\Omega)}.$$

Since $z_{u_\alpha - \bar{u}} \to 0$ in $L^2(\Omega)$, the following inequality is satisfied for all α small enough

$$\frac{\tau}{4}\|z_{u_\alpha - \bar{u}}\|_{L^2(\Omega)} + \frac{c_A}{4}\|u_\alpha - \bar{u}\|^{1+\frac{1}{\kappa}}_{L^1(A)} + \alpha\|u_\alpha - \bar{u}\|^2_{L^2(\Omega)} \le \alpha^2\|w\|^2_{L^2(\Omega)} + C\alpha^{\kappa+1},$$

which implies the claim for the second case. $\qquad\square$

CHAPTER 10

Conclusion and Outlook

We successfully analysed and tested three different regularization methods for optimal control problems governed by linear and semilinear partial differential equations. Additional constraints consists of box constraints for the control as well as inequality constraint for the state. Let us recall the results of this thesis and give some ideas for possible future research.

10.1   Linear State Equation

If the solution operator S of the underlying partial differential equation is linear we analysed two different regularization methods, depending on the additional constraints.

10.1.1   Control Constraints

The iterative Bregman method is suited for problems with additional control constraints, e.g.

$$\text{Minimize} \quad \frac{1}{2}\|Su - z\|_Y^2 \qquad (P_1)$$
$$\text{such that} \quad u_a \le u \le u_b \quad \text{a.e. in } \Omega.$$

This method reduces to the proximal point method for a specific choice of the regularization functional in the Bregman distance. However, for the proximal point method it is not clear how to obtain regularization error estimates under our regularity assumption (ASC).

If the solution of the optimal control problem (P_1) is unique, we proved that the iterative Bregman method is converging to this solution. In general we obtain that weak limit points of the sequence generated by the Bregman method are solutions to (P_1) .

If a solution of (P_1) satisfies the regularity assumption (ASC), which includes bang-bang solutions and non-attainability, we established regularization error estimates. These results generalize and extend known estimates for the PPM and Bregman iteration.

In the case of noisy data we presented an a-priori parameter choice rule and showed convergence results independent from a reachability condition. Furthermore we analysed the discretization error and its accumulation during the iterations. There we also allowed some inexactness for the solutions of the subproblems. This analysis revealed that the iterative Bregman method is robust and stable under the presence of numerical errors. This theoretical finding is supported by several numerical examples.

The analysis of this algorithm is heavily based on the linearity of the operator S and the absence of additional state constraints. We assume that most of the techniques can be carried over if additional state constraints are present. Here we have in our mind the regularization ideas of the agumented Lagrange method established in Chapter 6, but instead of applying a Tikhonov regularization, we apply a Bregman regularization.

Another interesting topic to analyse is the extension of the Bregman method to nonlinear solution operators S. In this case, several new analytical methods have to be developed, first to analyse the method and second to obtain regularization error estimates. A possible approach would be the linearization of the state equation in the current iterate. This would be in the spirit of some Newton-type methods.

10.1.2 Control and State Constraints

Next, we analysed an augmented Lagrange method for problems of the following form

$$\min \ J(y,u) := \frac{1}{2}\|y - y_d\|_{L^2(\Omega)}^2 + \beta\|u\|_{L^1(\Omega)}, \qquad (P_2)$$

subject to

$$
\begin{aligned}
Ay &= u & &\text{in } \Omega, \\
y &= 0 & &\text{on } \partial\Omega, \\
y &\leq \psi & &\text{in } \Omega, \\
u_a &\leq u \leq u_b & &\text{in } \Omega.
\end{aligned}
$$

Recall that A is a linear elliptic operator. Due to the absence of an additional L^2-regularization term in (P_2) this problem is numerically challenging. We combined an augmented Lagrange method to eliminate the state constraints and a Tikhonov regularization approach to solve (P_2). We presented an update rule for the multiplier and a coupling between the regularization and the penalty parameter. The key idea is to reduce the regularization parameter during the algorithm in a reasonable manner.

Due to this choice we obtain strong convergence of our algorithm. The arising problems are solved by an active-set method. Several numerical examples are shown to support our algorithm.

Although we established (strong) convergence of our algorithm we do not have regularization error estimates at hand. Here it would be interesting if the regularity assumption (ASC) can be used to derive such estimates.

10.2 Nonlinear State Equation

In the second part of this thesis we considered a non-linear state equation and optimal control problems of the form

$$\text{Minimize} \quad \frac{1}{2}\|y_u - y_d\|_{L^2(\Omega)}^2 + \beta\|u\|_{L^1(\Omega)} \tag{P_3}$$

$$\text{such that} \quad u_a \leq u \leq u_b \quad \text{a.e. in } \Omega,$$

with $\beta \geq 0$. Here y_u is the solution of the semi-linear Dirichlet problem

$$Ay + f(x, y) = u \quad \text{in } \Omega,$$
$$y = 0 \quad \text{on } \partial\Omega.$$

The non-linearity of the partial differential equation significantly increases the complexity of the problem. We analysed a Tikhonov regularization of problem (P_3). With the help of a sufficient second order condition and the regularity assumption (ASC$_\beta$) we derived regularization error estimates of the type

$$\|u_\alpha - \bar{u}\|_{L^2(\Omega)} = \mathcal{O}\left(\alpha^{d/2}\right) \text{ for } \alpha \to 0.$$

Note that we used different second order conditions for the case $\beta = 0$ and $\beta > 0$.

In addition we showed that in the case $\beta = 0$, the regularity assumption is not only sufficient for high regularization error rates, but also necessary. This should also carry over to the case $\beta > 0$.

The analysis presented in the section can be extended in future research to incorporate discretization errors, as was done in [28, 94, 96, 97] for linear-quadratic problems.

In future work it would also be interesting to analyse problem (P_3) with additional state constraints. This could be done in the spirit of Chapter 6. However, the nonlinearity increases the complexity, but the problem is still accessible with the augmented Lagrange method, see [55].

Bibliography

[1] A. Benfenati and V. Ruggiero. Inexact Bregman iteration with an application to Poisson data reconstruction. *Inverse Problems*, 29(6):065016, 31, 2013.

[2] M. Bergounioux. Augmented Lagrangian method for distributed optimal control problems with state constraints. *J. Optim. Theory Appl.*, 78(3):493–521, 1993.

[3] M. Bergounioux. On boundary state constrained control problems. *Numer. Funct. Anal. Optim.*, 14(5-6):515–543, 1993.

[4] M. Bergounioux, K. Ito, and K. Kunisch. Primal-dual strategy for constrained optimal control problems. *SIAM J. Control Optim.*, 37(4):1176–1194, 1999.

[5] M. Bergounioux and K. Kunisch. Primal-dual strategy for state-constrained optimal control problems. *Comput. Optim. Appl.*, 22(2):193–224, 2002.

[6] S. Beuchler, C. Pechstein, and D. Wachsmuth. Boundary concentrated finite elements for optimal boundary control problems of elliptic PDEs. *Comput. Optim. Appl.*, 51(2):883–908, 2012.

[7] J. F. Bonnans. Second-order analysis for control constrained optimal control problems of semilinear elliptic systems. *Appl. Math. Optim.*, 38(3):303–325, 1998.

[8] J. F. Bonnans and A. Shapiro. *Perturbation analysis of optimization problems*. Springer Series in Operations Research. Springer-Verlag, New York, 2000.

[9] L. M. Bregman. The relaxation method of finding the common point of convex sets and its application to the solution of problems in convex programming. *Ussr Computational Mathematics and Mathematical Physics*, 7:200–217, 1967.

[10] M. Burger. Bregman distances in inverse problems and partial differential equations. In *Advances in mathematical modeling, optimization and optimal control*, volume 109 of *Springer Optim. Appl.*, pages 3–33. Springer, [Cham], 2016.

[11] M. Burger, E. Resmerita, and L. He. Error estimation for Bregman iterations and inverse scale space methods in image restoration. *Computing*, 81(2–3):109–135, 2007.

[12] E. Casas. Control of an elliptic problem with pointwise state constraints. *SIAM J. Control Optim.*, 24(6):1309–1318, 1986.

[13] E. Casas. Second order analysis for bang-bang control problems of PDEs. *SIAM J. Control Optim.*, 50(4):2355–2372, 2012.

[14] E. Casas, C. Clason, and K. Kunisch. Approximation of elliptic control problems in measure spaces with sparse solutions. *SIAM J. Control Optim.*, 50(4):1735–1752, 2012.

[15] E. Casas, J. C. de los Reyes, and F. Tröltzsch. Sufficient second-order optimality conditions for semilinear control problems with pointwise state constraints. *SIAM J. Optim.*, 19(2):616–643, 2008.

[16] E. Casas, R. Herzog, and G. Wachsmuth. Optimality conditions and error analysis of semilinear elliptic control problems with L^1 cost functional. *SIAM J. Optim.*, 22(3):795–820, 2012.

[17] E. Casas and F. Tröltzsch. Second order analysis for optimal control problems: improving results expected from abstract theory. *SIAM J. Optim.*, 22(1):261–279, 2012.

[18] E. Casas and F. Tröltzsch. Second-order and stability analysis for state-constrained elliptic optimal control problems with sparse controls. *SIAM J. Control Optim.*, 52(2):1010–1033, 2014.

[19] E. Casas, F. Tröltzsch, and A. Unger. Second order sufficient optimality conditions for a class of elliptic control problems. In *Control and partial differential equations (Marseille-Luminy, 1997)*, volume 4 of *ESAIM Proc.*, pages 285–300. Soc. Math. Appl. Indust., Paris, 1998.

[20] E. Casas, D. Wachsmuth, and G. Wachsmuth. Sufficient second-order conditions for bang-bang control problems. *SIAM J. Control Optim.*, 55(5):3066–3090, 2017.

[21] Y. Censor and A. Lent. An iterative row-action method for interval convex programming. *J. Optim. Theory Appl.*, 34(3):321–353, 1981.

[22] Y. Censor and S. A. Zenios. Proximal minimization algorithm with D-functions. *J. Optim. Theory Appl.*, 73(3):451–464, 1992.

[23] Y. A. Censor and S. A. Zenios. *Parallel Optimization: Theory, Algorithms and Applications*. Oxford University Press, 1997.

[24] G. Chavent and K. Kunisch. Convergence of Tikhonov regularization for constrained ill-posed inverse problems. *Inverse Problems*, 10(1):63–76, 1994.

[25] G. Chen and M. Teboulle. Convergence analysis of a proximal-like minimization algorithm using Bregman functions. *SIAM J. Optim.*, 3(3):538–543, 1993.

[26] F. H. Clarke. A new approach to Lagrange multipliers. *Math. Oper. Res.*, 1(2):165–174, 1976.

[27] J. C. De los Reyes. *Numerical PDE-constrained optimization.* Springer, Heidelberg, 2015.

[28] K. Deckelnick and M. Hinze. A note on the approximation of elliptic control problems with bang-bang controls. *Comput. Optim. Appl.*, 51(2):931–939, 2012.

[29] M. Dobrowolski. *Angewandte Funktionalanalysis.* Springer-Verlag, Berlin, 2006.

[30] J. Eckstein. Nonlinear proximal point algorithms using Bregman functions, with applications to convex programming. *Math. Oper. Res.*, 18(1):202–226, 1993.

[31] J. Eckstein. Approximate iterations in Bregman-function-based proximal algorithms. *Math. Programming*, 83(1, Ser. A):113–123, 1998.

[32] J. Eckstein and D. P. Bertsekas. On the Douglas-Rachford splitting method and the proximal point algorithm for maximal monotone operators. *Math. Programming*, 55(3, Ser. A):293–318, 1992.

[33] H. W. Engl, M. Hanke, and A. Neubauer. *Regularization of inverse problems*, volume 375 of *Mathematics and its Applications*. Kluwer Academic Publishers Group, Dordrecht, 1996.

[34] F. Facchinei and J.-S. Pang. *Finite-dimensional variational inequalities and complementarity problems. Vol. I.* Springer Series in Operations Research. Springer-Verlag, New York, 2003.

[35] P. E. Farrell, D. A. Ham, S. W. Funke, and M. E. Rognes. Automated derivation of the adjoint of high-level transient finite element programs. *SIAM J. Sci. Comput.*, 35(4):C369–C393, 2013.

[36] U. Felgenhauer. On stability of bang-bang type controls. *SIAM J. Control Optim.*, 41(6):1843–1867, 2003.

[37] K. Frick and M. Grasmair. Regularization of linear ill-posed problems by the augmented Lagrangian method and variational inequalities. *Inverse Problems*, 28(10):104005, 16, 2012.

[38] K. Frick, D.A. Lorenz, and E. Resmerita. Morozov's principle for the augmented Lagrangian method applied to linear inverse problems. *Multiscale Model. Simul.*, 9(4):1528–1548, 2011.

[39] K. Frick and O. Scherzer. Regularization of ill-posed linear equations by the non-stationary augmented Lagrangian method. *J. Integral Equations Appl.*, 22(2):217–257, 2010.

[40] S. W. Funke and P. E. Farrell. A framework for automated pde-constrained optimisation. *CoRR*, abs/1302.3894, 2013.

[41] T. Goldstein and S. Osher. The split Bregman method for $L1$-regularized problems. *SIAM J. Imaging Sci.*, 2(2):323–343, 2009.

[42] O. Güler. On the convergence of the proximal point algorithm for convex minimization. *SIAM J. Control Optim.*, 29(2):403–419, 1991.

[43] J. Hadamard. *Lectures on Cauchy's problem in linear partial differential equations*. Dover Publications, New York, 1953.

[44] P. Hájek and M. Johanis. *Smooth analysis in Banach spaces*, volume 19 of *De Gruyter Series in Nonlinear Analysis and Applications*. De Gruyter, Berlin, 2014.

[45] M. Hanke and C. W. Groetsch. Nonstationary iterated Tikhonov regularization. *J. Optim. Theory Appl.*, 98(1):37–53, 1998.

[46] M. Hinze. A variational discretization concept in control constrained optimization: The linear-quadratic case. *Computational Optimization and Applications*, 30(1):45–61, 2005.

[47] M. Hinze and U. Matthes. A note on variational discretization of elliptic Neumann boundary control. *Control Cybernet.*, 38(3):577–591, 2009.

[48] M. Hinze, R. Pinnau, M. Ulbrich, and S. Ulbrich. *Optimization with PDE constraints*, volume 23 of *Mathematical Modelling: Theory and Applications*. Springer, New York, 2009.

[49] K. Ito and B. Jin. A new approach to nonlinear constrained Tikhonov regularization. *Inverse Problems*, 27(10):105005, 23, 2011.

[50] K. Ito and K. Kunisch. The augmented Lagrangian method for equality and inequality constraints in Hilbert spaces. *Math. Programming*, 46(3, (Ser. A)):341–360, 1990.

[51] K. Ito and K. Kunisch. Semi-smooth Newton methods for state-constrained optimal control problems. *Systems Control Lett.*, 50(3):221–228, 2003.

[52] K. Ito and K. Kunisch. Semi-smooth Newton methods for variational inequalities of the first kind. *M2AN Math. Model. Numer. Anal.*, 37(1):41–62, 2003.

[53] M. Jaćimović, I. Krnić, and O. Obradović. On the convergence of a class of the regularization methods for ill-posed quadratic minimization problems with constraint. *Publ. Inst. Math. (Beograd) (N.S.)*, 97(111):89–102, 2015.

[54] J. Jost and X. Li-Jost. *Calculus of variations*, volume 64 of *Cambridge Studies in Advanced Mathematics*. Cambridge University Press, Cambridge, 1998.

[55] C. Kanzow, D. Steck, and D. Wachsmuth. An augmented Lagrangian method for optimization problems in Banach spaces. *SIAM J. Control Optim.*, to appear.

[56] A. Kaplan and R. Tichatschke. *Stable methods for ill-posed variational problems*, volume 3 of *Mathematical Topics*. Akademie Verlag, Berlin, 1994. Prox-regularization of elliptic variational inequalities and semi-infinite problems.

[57] A. Kaplan and R. Tichatschke. On inexact generalized proximal methods with a weakened error tolerance criterion. *Optimization*, 53(1):3–17, 2004.

[58] V. Karl and F. Pörner. A Joint Tikhonov Regularization and Augmented Lagrange Approach for Ill-posed State Constrained Control Problems with Sparse Controls. *to be published in Numerical Functional Analysis and Optimization*, 2018.

[59] V. Karl and D. Wachsmuth. An augmented Lagrange method for elliptic state constrained optimal control problems. *Computational Optimization and Applications*, Nov 2017.

[60] K. C. Kiwiel. Proximal minimization methods with generalized Bregman functions. *SIAM J. Control Optim.*, 35(4):1142–1168, 1997.

[61] K. Kohls, A. Rösch, and K. G. Siebert. A posteriori error analysis of optimal control problems with control constraints. *SIAM J. Control Optim.*, 52(3):1832–1861, 2014.

[62] K. Krumbiegel, I. Neitzel, and A. Rösch. Regularization for semilinear elliptic optimal control problems with pointwise state and control constraints. *Comput. Optim. Appl.*, 52(1):181–207, 2012.

[63] F. Liu and M. Z. Nashed. Regularization of nonlinear ill-posed variational inequalities and convergence rates. *Set-Valued Anal.*, 6(4):313–344, 1998.

[64] A. Logg, K.-A. Mardal, G. N. Wells, et al. *Automated Solution of Differential Equations by the Finite Element Method*. Springer, 2012.

[65] A. Logg and G. N. Wells. Dolfin: Automated finite element computing. *ACM Transactions on Mathematical Software*, 37(2), 2010.

[66] D. A. Lorenz and A. Rösch. Error estimates for joint Tikhonov and Lavrentiev regularization of constrained control problems. *Appl. Anal.*, 89(11):1679–1691, 2010.

[67] B. Martinet. Brève communication. régularisation d'inéquations variationnelles par approximations successives. *ESAIM: Mathematical Modelling and Numerical Analysis - Modélisation Mathématique et Analyse Numérique*, 4(R3):154–158, 1970.

[68] A. A. Milyutin and N. P. Osmolovskiĭ. *Calculus of variations and optimal control*, volume 180 of *Translations of Mathematical Monographs*. American Mathematical Society, Providence, RI, 1998. Translated from the Russian manuscript by Dimitrii Chibisov.

[69] James D. Murray. *Mathematical Biology II: Spatial Models and Biomedical Applications (Interdisciplinary Applied Mathematics) (v. 2)*. Springer, 2011.

[70] A. Neubauer. Tikhonov-regularization of ill-posed linear operator equations on closed convex sets. *J. Approx. Theory*, 53(3):304–320, 1988.

[71] A. Neubauer. Tikhonov regularisation for nonlinear ill-posed problems: optimal convergence rates and finite-dimensional approximation. *Inverse Problems*, 5(4):541–557, 1989.

[72] S. Osher, M. Burger, D. Goldfarb, J. Xu, and W. Yin. An iterative regularization method for total variation-based image restoration. *Multiscale Model. Simul.*, 4(2):460–489 (electronic), 2005.

[73] N. P. Osmolovskiĭ. Quadratic conditions for nonsingular extremals in optimal control (a theoretical treatment). *Russian J. Math. Phys.*, 2(4):487–516 (1995), 1994.

[74] F. Pörner. A sharp regularization error estimate for bang-bang solutions for an iterative Bregman regularization method for optimal control problems. *PAMM*, 16(1):787–788, 2016.

[75] F. Pörner. Inexact iterative Bregman method for optimal control problems. *Numerical Functional Analysis and Optimization*, 2017.

[76] F. Pörner. A priori stopping rule for an iterative Bregman method for optimal control problems. *Optimization Methods and Software*, 33(2):249–267, 2018.

[77] F. Pörner and D. Wachsmuth. An iterative Bregman regularization method for optimal control problems with inequality constraints. *Optimization*, 65(12):2195–2215, 2016.

[78] F. Pörner and D. Wachsmuth. Tikhonov regularization of optimal control problems governed by semi-linear partial differential equations. *Math. Control Relat. Fields*, 8(1):315–335, 2018.

[79] S. Reich and S. Sabach. A strong convergence theorem for a proximal-type algorithm in reflexive Banach spaces. *J. Nonlinear Convex Anal.*, 10(3):471–485, 2009.

[80] S. Reich and S. Sabach. Two strong convergence theorems for a proximal method in reflexive Banach spaces. *Numer. Funct. Anal. Optim.*, 31(1–3):22–44, 2010.

[81] S. Reich and S. Sabach. Two strong convergence theorems for Bregman strongly nonexpansive operators in reflexive Banach spaces. *Nonlinear Anal.*, 73(1):122–135, 2010.

[82] R. T. Rockafellar. Augmented Lagrangians and applications of the proximal point algorithm in convex programming. *Math. Oper. Res.*, 1(2):97–116, 1976.

[83] S. Rotin. *Konvergenz des Proximal-Punkt-Verfahrens für inkorrekt gestellte Optimalsteuerprobleme mit partiellen Differentialgleichungen*. PhD thesis, Universität Trier, 1999.

[84] W. Rudin. *Functional analysis*. International Series in Pure and Applied Mathematics. McGraw-Hill, Inc., New York, second edition, 1991.

[85] M. Růžička. *Nichtlineare Funktionalanalysis*. Springer-Verlag, 2004.

[86] W. Schirotzek. *Nonsmooth analysis*. Universitext. Springer, Berlin, 2007.

[87] M. Seydenschwanz. Convergence results for the discrete regularization of linear-quadratic control problems with bang-bang solutions. *Comput. Optim. Appl.*, 61(3):731–760, 2015.

[88] M. V. Solodov and B. F. Svaiter. Forcing strong convergence of proximal point iterations in a Hilbert space. *Math. Program.*, 87(1, Ser. A):189–202, 2000.

[89] G. Stadler. Elliptic optimal control problems with L^1-control cost and applications for the placement of control devices. *Comput. Optim. Appl.*, 44(2):159–181, 2009.

[90] G. Stampacchia. Le problème de Dirichlet pour les équations elliptiques du second ordre à coefficients discontinus. *Ann. Inst. Fourier (Grenoble)*, 15(fasc. 1):189–258, 1965.

[91] N. Thanh Qui and D. Wachsmuth. Full Stability for a Class of Control Problems of Semilinear Elliptic Partial Differential Equations. *ArXiv e-prints*, October 2017.

[92] N. Thanh Qui and D. Wachsmuth. Stability for Bang-Bang Control Problems of Partial Differential Equations. *ArXiv e-prints*, July 2017.

[93] F. Tröltzsch. *Optimal control of partial differential equations*, volume 112 of *Graduate Studies in Mathematics*. American Mathematical Society, Providence, RI, 2010. Theory, methods and applications, Translated from the 2005 German original by Jürgen Sprekels.

[94] N. von Daniels. *Bang-bang control of parabolic equations*. PhD thesis, University of Hamburg, 2016.

[95] N. von Daniels. Tikhonov regularization of control-constrained optimal control problems, 2017.

[96] D. Wachsmuth. Adaptive regularization and discretization of bang-bang optimal control problems. *Electron. Trans. Numer. Anal.*, 40:249–267, 2013.

[97] D. Wachsmuth and G. Wachsmuth. Regularization error estimates and discrepancy principle for optimal control problems with inequality constraints. *Control Cybernet.*, 40(4):1125–1158, 2011.

[98] D. Wachsmuth and G. Wachsmuth. Necessary conditions for convergence rates of regularizations of optimal control problems. In *System modeling and optimization*, volume 391 of *IFIP Adv. Inf. Commun. Technol.*, pages 145–154. Springer, Heidelberg, 2013.

[99] G. Wachsmuth and D. Wachsmuth. Convergence and regularization results for optimal control problems with sparsity functional. *ESAIM Control Optim. Calc. Var.*, 17(3):858–886, 2011.

[100] D. Werner. *Funktionalanalysis*. Springer-Verlag, Berlin, extended edition, 2000.

[101] W. Yin, S. Osher, D. Goldfarb, and J. Darbon. Bregman iterative algorithms for l_1-minimization with applications to compressed sensing. *SIAM J. Imaging Sci.*, 1(1):143–168, 2008.

www.ingramcontent.com/pod-product-compliance
Lightning Source LLC
Chambersburg PA
CBHW061816210326
41599CB00034B/7018